# REAL-TIME RF-DNA FINGERPRINTING OF ZIGBEE DEVICES

# USING A SOFTWARE-DEFINED RADIO WITH FPGA PROCESSING

## I. Introduction

### 1.1 Introduction

This chapter provides the operational and technical motivations for conducting this research. Section 1.2 describes the Operational Motivation for focusing on ZigBee wireless network applications. Section 1.3 provides the Technical Motivation which is based on prior Air Force Institute of Technology (AFIT) Radio Frequency Distinct Native Attribute (RF-DNA) fingerprinting work, and the relative contributions of this research. Section 1.4 provides organizational details for this document.

### 1.2 Operational Motivation

Wireless Personal Area Networks (WPANs) are increasingly popular in personal, medical, industrial and other applications. The Institute of Electrical and Electronics Engineers (IEEE) 802.15.4 standard provides a specification for wireless mesh networks on which the ZigBee protocol is based. By design, ZigBee devices able to form WPANs where low-cost and extended battery life are desirable features. Traditional security techniques for ZigBee networks are predominantly based on presenting and verifying device bit-level credentials (e.g. keys). While historically effective, ZigBee networks remain vulnerable to attack by unauthorized rogue devices that can obtain and present false bit-level credentials matching an authorized device. Even without prior knowledge of the correct key, replay attacks against inadequately-defended networks can still be employed in which a packet transmitted by an authorized device is collected and later replayed by an unauthorized device [18].

## 1.3 Technical Motivation

As shown in Table 1.1 there is a considerable amount of previous related research [4,12,13,17,20,29,32,33,34] addressing Physical (PHY) layer of security of wireless communication systems. Some of the methods in these works were adopted and applied here to address ZigBee PHY-based bit-level security augmentation. The additional PHY security is achieved using Radio Frequency Distinct Native Attribute (RF-DNA) fingerprinting. RF-DNA exploitation involves generating device "fingerprint" from PHY waveform responses to achieve human-like device discrimination–a unique one-to-one association between a fingerprint and a device. The RF-DNA fingerprint used to discriminate among devices, even when identical bit-level credentials are presented. While previous AFIT research has demonstrated the effectiveness of MATLAB simulation-based RF-DNA classification of ZigBee devices [5,20,23], the research here represents the next step towards achieving real-time device classification and verification. A complete RF-DNA based security solution for ZigBee devices in the form of an air monitor is proposed in [23]. The air monitor would be physically co-located with ZigBee devices and actively accept or reject signals from other ZigBee devices based on their fingerprint signatures. The purpose of this research was to demonstrate feasibility of implementing an air monitor using a National Instruments (NI) X310 Software-Defined Radio (SDR) hosting a Kintex-7 Field Programmable Gate Array (FPGA).

Table 1.1 provides a summary of technical areas that were previously addressed and areas addressed in this research. The amount of previous related research listed in Table 1.1 shows that the efficacy of RF-DNA fingerprinting has been well-established. The general methodology of the RF-DNA fingerprinting process has remained relatively unchanged in this research given the success of these previous works.

## 1.4 Document Organization

The remainder of this document is organized as follows. Chapter 2 provides a basic outline of the ZigBee protocol, SDR implementation, and background on the RF-DNA processes employed in this research. Chapter 3 provides the methodology used for experimental signal collection, FPGA hardware design, classification, three fingerprint generation methods, device ID verification, and DRA. Chapter 4 presents classification results for the three fingerprint generation methods, classification performance using DRA feature sets, device ID verification, rogue rejection and FPGA resource utilization. Chapter 5 provides a summary and conclusions based on research results and recommendations for future work.

Table 1.1: Technical areas in previous related work and current research contributions.

| Technical Area | Addressed | Previous Work Ref # | This Work |
|---|---|---|---|
| TD Features | X | [4,13,17,20,29,32,33,34] | X |
| SD Features | X | [27,29,33] | |
| WD Features | X | [12,12] | |

| Fingerprint Generation Platform | | | |
|---|---|---|---|
| Computer | X | [4,10,11,12,13,17, 26,27,29,32,33,34] | X |
| FPGA | X | | X |

| Signal Type | | | |
|---|---|---|---|
| 802.11 WiFi | X | [11,12,33] | |
| GSM Cellular | X | [24] | |
| 802.16e WiMax | X | [26,27,32,33] | |
| 802.15.4 ZigBee | X | [4,20] | X |

| Classifier Type | | | |
|---|---|---|---|
| MDA/ML | X | [4,10,12,17,20,26,27,29,32,33,34] | X |
| GRLVQI | X | [4,11,17] | X |

| Dimensional Reduction Analysis (DRA) | | | |
|---|---|---|---|
| GRLVQI | X | [4,11,17] | X |
| LFS | X | [10] | |

## II. Background

This chapter provides the technical background supporting the methodology described in Chapter 3. Section 2.1 provides details for the ZigBee protocol defined by the IEEE 802.15.4 standard for Wireless Personal Area Networks (WPANs) [9]. Section 2.2 describes the Radio Frequency Distinct Native Attribute (RF-DNA) fingerprint generation process which includes calculation of statistical features over a selected Region Of Interest (ROI) within time-domain signal responses. Section 2.3 describes model development and device discrimination using the Multiple Discriminant Analysis, Maximum Likelihood (MDA/ML) classifier. This is followed by Section 2.4 which describes the Generalized Relevance Learning Vector Quantization-Improved (GRLVQI) classifier. Section 2.5 provides a description of Software-Defined Radio (SDR) implementation and benefits. The chapter concludes with Section 2.6 that describes attributes of a Field Programmable Gate Array (FPGA) and benefits for its use.

### 2.1 ZigBee Signal Characteristics

ZigBee devices are used to form low-cost, low-power WPANs and support network-enabled home appliances, home automation, industrial control, medical data monitoring and other applications. ZigBee devices are designed according to the IEEE 802.15.4 standard [9] which includes provisions supporting several possible modulation schemes and frequency bands. For this research, the IEEE 802.15.4 frequency band

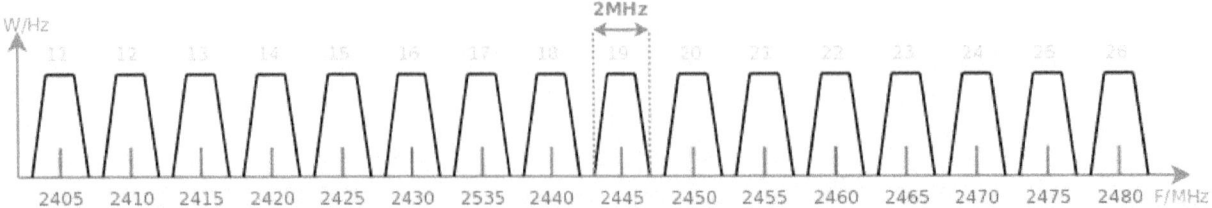

Figure 2.1: Spectral Location of ZigBee Channels Number 11-26 [28].

spanning 2400.0 to 2483.5 MHz was used with 16-ary Offset-Quadrature Phase Shift Keying (O-QPSK) data modulation. Each channel has an instantaneous RF bandwidth of $BW_{RF}$ = 2.0 MHz, with $\Delta_{Ch}$ =5.0 MHz spacing between adjacent channels. Figure 2.1 shows the spectral location and assignment of channels 11-26.

ZigBee transmissions are specified to begin with a preamble region consisting of 8 O-QPSK symbols mapping to 32 binary zeros. Previous research [13] has shown that this preamble region can be successfully exploited to generate fingerprints and provide reliable device discrimination using an MDA/ML classifier; this is described in greater detail in Section 2.2 and Section 2.3.

## 2.2 RF-DNA Fingerprint Generation

RF-DNA fingerprinting is the process of characterizing the inherent differences in emission responses collected from multiple devices. These differences are the result of factors such as operating temperature, device age, and variations in manufacturing tolerance [25]. Fingerprints can be generated from multiple responses, in multiple untransformed and transformed domains. Fingerprints in this work were generated using a two step process: 1) generation of instantaneous Time-Domain (TD) responses, and

2) statistical feature generation over the selected TD ROI. Each step is described in detail below.

### 2.2.1 Time-Domain Waveform Response

The common TD features used for RF-DNA fingerprint generation are instantaneous amplitude ($a$), phase ($\phi$), and frequency ($f$) responses of a given ROI. Elements of the corresponding discrete sequences $\{a[n]\}$, $\{\phi[n]\}$, and $\{f[n]\}$ are calculated from Real ($Re[n]$) and Imaginary ($Im[n]$) ROI components as follows [2,3,4]:

$$a[n] = \sqrt{Re[n]^2 + Im[n]^2}\,, \qquad (2.1)$$

$$\phi[n] = tan^{-1}\left|\frac{Im[n]}{Re[n]}\right|, \text{ for } Re[n] \neq 0\,, \qquad (2.2)$$

$$f[n] = \frac{1}{2\pi}\left|\frac{d\phi[n]}{dt}\right|. \qquad (2.3)$$

The resultant $\{a[n]\}$, $\{\phi[n]\}$, and $\{f[n]\}$ sequences are centered (mean removed) and normalized as follows:

$$\bar{a}_c[n] = \frac{a[n] - \mu_a}{\max_n\{a_c[n]\}}, \qquad (2.4)$$

$$\bar{\phi}_c[n] = \frac{\phi[n] - \mu_\phi}{\max_n\{\phi_c[n]\}}, \qquad (2.5)$$

$$\bar{f}_c[n] = \frac{f[n] - \mu_f}{\max_n\{f_c[n]\}}, \qquad (2.6)$$

where $\mu_a$, $\mu_\phi$, and $\mu_f$ are corresponding sequence means and "max" represents the maximum value of the centered sequences $\{a_c[n]\}$, $\{\phi_c[n]\}$ and $\{f_c[n]\}$. The resultant $\{\bar{a}_c[n]\}$, $\{\bar{\phi}_c[n]\}$ and $\{\bar{f}_c[n]\}$ sequences are the centered, normalized TD sequences used for statistical RF-DNA feature calculation.

### 2.2.2 Statistical RF-DNA Features

The centered, normalized sequences from (2.4) through (2.6) are divided into $N_R$ equal length, contiguous subsequences (ROI subregions) and summarized using $N_M = 4$ statistical metrics of standard deviation $(\sigma)$, variance $(\sigma^2)$, skewness $(\gamma)$ and kurtosis $(\kappa)$ that are computed over each of $N_R$ subregions. Each of the summary statistics is also computed over the entire ROI, resulting in a total of $N_R + 1$ regions being used. This process is illustrated in Figure 2.2.

The summary $\sigma, \sigma^2, \gamma$ and $\kappa$ statistics are calculated as follows [25]:

$$\sigma = \sqrt{\frac{1}{N} \sum_{n=1}^{N} (x[n] - \mu)^2}, \tag{2.7}$$

$$\sigma^2 = \frac{1}{N} \sum_{n=1}^{N} (x[n] - \mu)^2, \tag{2.8}$$

$$\gamma = \frac{1}{N\sigma^3} \sum_{n=1}^{N} (x[n] - \mu)^3, \tag{2.9}$$

$$\kappa = \frac{1}{N\sigma^4} \sum_{n=1}^{N} (x[n] - \mu)^4, \tag{2.10}$$

where $N$ is total number of samples in a given subsequence and $x[n]$ represents a an arbitrary TD response. The summary statistics for a given region $R_i$ are then concatenated to form vector $F_{R_i}$ as follows:

$$F_{R_i} = [\sigma_{R_i} \sigma_{R_i}^2 \gamma_{R_i} \kappa_{R_i}]_{1\times4}, \tag{2.11}$$

where $i = 1,2,3, \dots, N_R + 1$. The vector $\mathbf{F}_{R_i}$ is computed for each of $N_R + 1$ regions, and used to form vector $\mathbf{F}^x$ as,

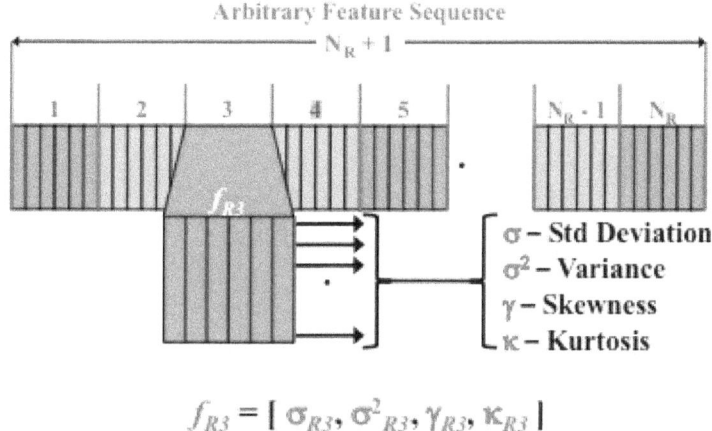

$$f_{R3} = [\ \sigma_{R3},\ \sigma^2_{R3},\ \gamma_{R3},\ \kappa_{R3}\ ]$$

Figure 2.2: Representative Illustration of RF-DNA Statistical Fingerprint Generation for $N_R + 1$ Total Subregions [25].

$$\mathbf{F}^x = [F_{R_1} \vdots F_{R_2} \vdots F_{R_3} \cdots F_{R_{N_R+1}}]_{1\times[N_M\times(N_R+1)\times N_C]}, \qquad (2.12)$$

where $x$ represents one of the TD responses $a$, $\phi$ or $f$. Finally, the composite statistical fingerprint vector $\mathbf{F}$ is formed by concatenating the $\mathbf{F}^x$ vector of each TD response as follows:

$$\mathrm{F} = [\mathrm{F}^a \vdots \mathrm{F}^\phi \vdots \mathrm{F}^f]_{1\times(N_R+1)\times3} \qquad (2.13)$$

The resultant full-dimensional fingerprint vector F from (2.13) contains a total of $N_f = (\# \text{ of TD Features}) \times (\# \text{ of Statistical Metrics}) \times (\# \text{ of Regions})$ elements.

## 2.3 MDA/ML Classification

MDA/ML device discrimination includes two distinct processes: Multiple Discriminant Analysis (MDA) model development and Maximum Likelihood (ML) classification. The goal of MDA is to reduce feature dimensionality and provide the greatest separation between multiple input classes. This is accomplished by projecting full-dimensional fingerprints into a lower-dimensional space, while 1) maximizing between-class spread and 2) minimizing within-class spread. The between-class ($S_b$) and within-class ($S_w$) scatter matrices are computed using [31]:

$$S_b = \sum_{i=1}^{N_C} P_i \, \Sigma_i, \tag{2.14}$$

$$S_w = \sum_{i=1}^{N_C} P_i (\mu_i - \mu_0)(\mu_i - \mu_0)^T, \tag{2.15}$$

where $\Sigma_i$ and $P_i$ are the covariance matrix and prior probability, respectively, for each of the $N_C$ input classes. The $N_f$-dimensional input RF-DNA fingerprint vectors, $\vec{F}$ from (2.13), are then projected into the $N_C - 1$-dimensional space according to

$$\hat{f} = \mathbf{W}^T \vec{F}, \tag{2.16}$$

where $\mathbf{W}$ is the $N_f \times (N_C - 1)$ projection matrix formed from the $N_C - 1$ Eigenvectors of $S_w^{-1} S_b$ and $\hat{f}$ is the RF-DNA fingerprint projection into the lower dimensional subspace.

Classification performance depends on the effectiveness of the $\mathbf{W}$ matrix in maximizing between-class distance and minimizing within-class spread. To illustrate the projection of $\vec{F}$ using $\mathbf{W}$, two possible MDA projection matrices ($\mathbf{W_1}$ and $\mathbf{W_2}$) are shown in Figure 2.3 [25]. In this case, projection matrix $\mathbf{W_1}$ provides the "best" classification model.

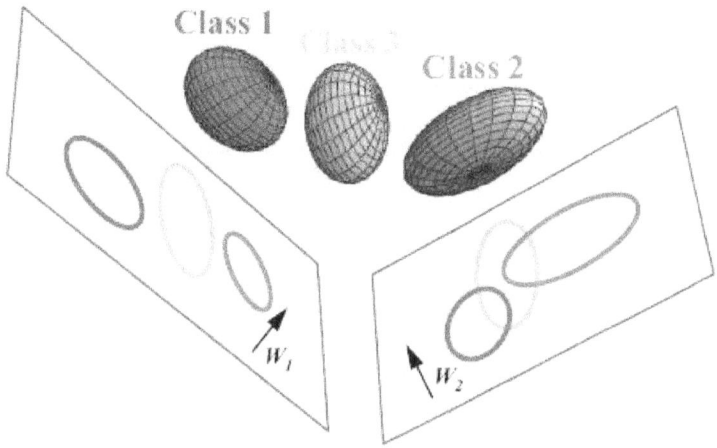

Figure 2.3: Representative Projections using $W_1$ and $W_2$ for a $N_C = 3$ Class Problem into a 2-Dimensional Space [25].

Classification of the projected fingerprints $\hat{f}$ is performed using a Maximum Likelihood (ML) process based on Bayesian posterior probability and assuming uniform costs and equal prior probability. A similarity measure is computed by comparing the likeness of the unknown $\hat{f}$ fingerprint to each of $N_C = 3$ classes. The classification decision is made by assigning (rightly or wrongly) unknown fingerprint $\hat{f}$ to the class yielding the highest measure of similarity.

## 2.4 GRLVQI Classification

The GRLVQI classifier was also considered to provide a comparison to MDA/ML. Like MDA/ML, GRLVQI is used to discriminate between multiple classes but provides several advantages, namely: 1) there is no inherent assumption of the distribution of input data, 2) GRLVQI is more suitable for cases where input class data (fingerprints) is noisy or inconsistent, and 3) a relevance ranking is generated for each

RF-DNA fingerprint feature [25]. Like prior related research, the GRLVQI relevance ranking is of particular interest given it provides a means numerically rank features and enable Dimensional Reduction Analysis (DRA).

For this research, GRLVQI was implemented as described in [25], with $N_p = 20$ prototype vectors representing each of the $N_C$ classes. An RF-DNA fingerprint is classified as one of $N_C$ classes by measuring the Euclidean distance between mapped fingerprints and each prototype vector; the input fingerprint is assigned to the class whose prototype vector is the minimum Euclidean distance from the mapped fingerprint. Figure 2.4 displays a visualization of the GRLVQI classification process.

## 2.5 Software-Defined Radio (SDR)

A Software-Defined Radio (SDR) is a radio system in which components that have been traditionally implemented with analog hardware, such as mixers or filters, are replaced with a software-based implementation. SDR technology has quickly gaining popularity within the last decade given the increased performance in embedded microprocessors and general-purpose computers which enable implementation of highly complex radio systems. An SDR can be rapidly reconfigured to change modulation scheme, bandwidth, and other key parameters that are normally fixed in an analog design. The SDR functionality can be implemented in a general purpose computer, Field Programmable Gate Array (FPGA), or any combination thereof. Implementations that only rely on a general purpose computer can have extremely high latency because the signal has to propagate from the receiver to the computer system, normally through a Universal Serial Bus (USB) or Ethernet connection. Components implemented on an

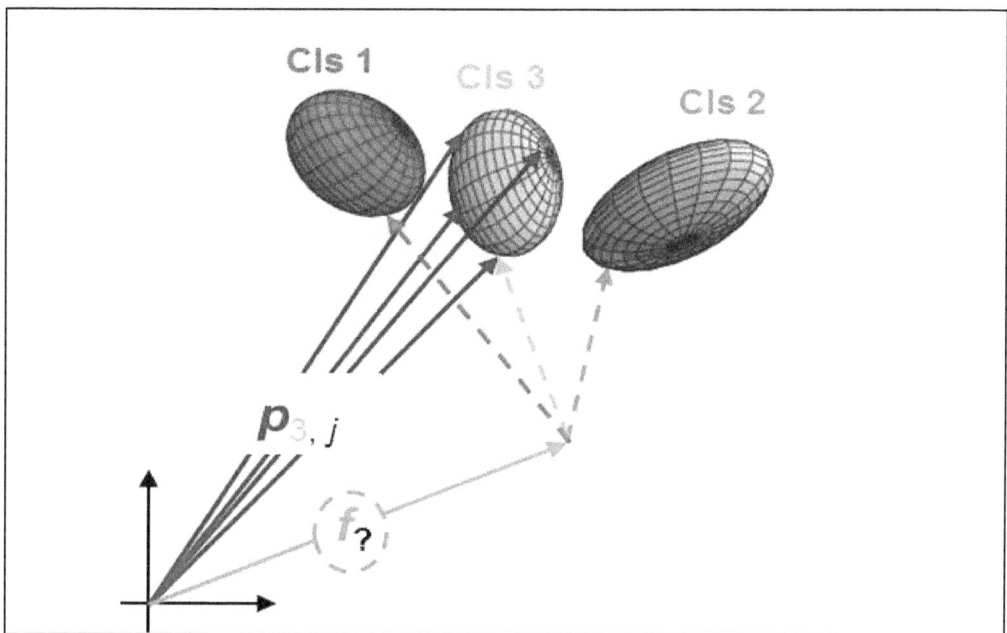

Figure 2.4: GRLVQI Classification of an Unknown Fingerprint Based on Minimum Euclidean Distance [25].

FPGA can have very low latency, achieving speeds close to that of an integrated circuit specifically designed for the task at hand.

## 2.6 Field Programmable Gate Array (FPGA)

A Field Programmable Gate Array (FPGA) is a chip that can be programmed to rearrange its internal logic gates to perform a specific function. For example, an FPGA can be programmed to process images, implement an encryption algorithm or even to act as a general purpose microprocessor. An FPGA is typically chosen in an application where specialized operations are required and where requirements are expected to change over time. In an SDR system as described in Section 2.5, an FPGA can take the place of mixers, filters, and more. These operations that were once performed with fixed hardware can be rapidly reconfigured in an FPGA implementation. Additionally, the

FPGA can be physically located in close proximity to an Analog-to-Digital Converter (A/D) to minimize overall latency of the radio system. Because of the increasing capabilities of modern FPGAs, more SDR functionality can be migrated toward FPGA implementation and away from general purpose computers.

### 2.6.1 Coordinate Rotation Digital Computer (CORDIC)

Many communications systems require computation of instantaneous phase as part of the modulation/demodulation scheme. Instantaneous phase is calculated using an arctangent operation as shown in (2.2). There are multiple algorithms available to implement trigonometric function using an FPGA, Coordinate Rotation Digital Computer (CORDIC) being one of the most popular. CORDIC is a hardware-efficient algorithm that only requires addition, subtraction and table lookup operations [2]. The CORDIC can operate in two modes: vectoring mode and rotation mode. When operating in rotation mode, the sine and cosine values for a given angle are calculated. This operation is accomplished by rotating a unit length vector from the x-axis to the given angle with successively smaller rotations until the given angle is reached. The sine and cosine values are determined based on the direction of each angular rotation. Using this technique, a vector can be converted from a polar to rectangular coordinate system. The rotation mode of the CORDIC process is illustrated in Figure 2.5.

When operating in vectoring mode, the CORDIC operations described above are reversed and the magnitude and angle are generated for a given x-y coordinate pair. Vectoring mode operation results in a conversion from a rectangular to polar coordinate system.

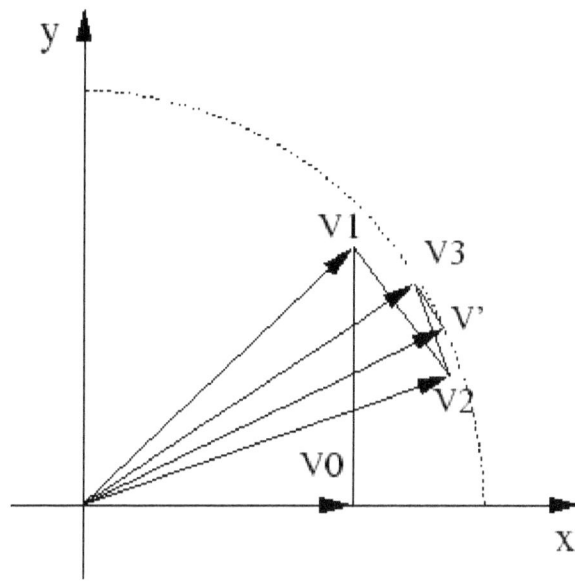

Figure 2.5: Example Operation of CORDIC in Rotation Mode [14].

### 2.6.2 Cascaded Integrator-Comb (CIC) Filter

In a high-speed SDR system, extracting a narrow-band signal requires down-sampling and filtering. For high decimation rates, a long filter with many coefficients is required for sufficient anti-aliasing filtering. This can become a large bottleneck in the SDR system, both in terms of latency time and required hardware resources [3].

The Cascaded Integrator-Comb (CIC) filter is a popular filter choice for SDR systems [16]. The CIC filter operates using only addition and subtraction operations; there is no multiplication required in the CIC design. This makes the CIC a particularly attractive option for FPGA-based SDR systems where hardware resources are limited and multiply operations are especially costly in terms of FPGA resources. Additionally, unlike most discrete filters the, CIC has a decimator built into its architecture.

A CIC design consists of $n_s$ "integrator" addition stages followed by the same number of "comb" subtraction stages. The decimator can be located either at the end of

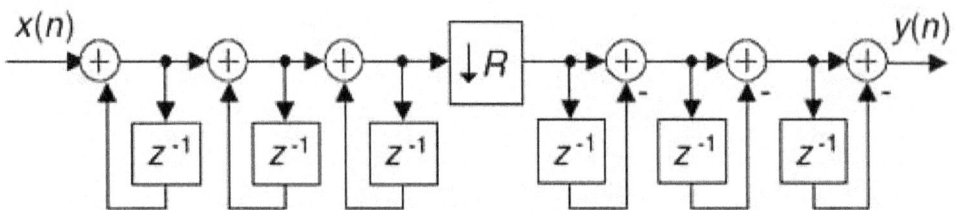

Figure 2.6: Example CIC Design with $n_s = 3$ Stages [15]. The $\oplus$ Symbols Denote Modulo-2 Addition, $Z^1$ Denotes a Delay by One Clock Cycle, and $\downarrow R$ Denotes Down-Sampling (Decimation).

all stages or between the integrator and comb stages. An example CIC design is shown

in Figure 2.6. In the case of Figure 2.6, the decimator is located between the integrator

and comb stages.

## III. Methodology

Wireless communication devices are inherently insecure because the transmission medium can be accessed by unauthorized users. Traditional mechanisms to secure the communications channel are based on encoding information at the bit level only. These security mechanisms can be bypassed by forging the required bit-level credentials. This research aims to characterize a hardware-based security mechanism for protecting wireless systems from malicious attacks. The proposed solution generates real-time Radio Frequency Distinct Native Attribute (RF-DNA) fingerprints as described in Chapter 2.

This chapter describes the methodology used to obtain the experimental results presented in Chapter 4. The experimental X310 SDR methodology for assessing RF-DNA fingerprinting in this research is shown in Figure 3.1.

A simplified fingerprint generation scheme suitable for implementation on a Field Programmable Gate Array (FPGA) was developed. The signal of interest for this research was ZigBee emissions. A MATLAB model was created to validate the performance of the new prototypical fingerprint generation process. ZigBee beacon requests (bursts) were experimentally collected on the X310 Software-Defined Radio (SDR) platform. The collected signals were processed in MATLAB to generate $\vec{F}_M$ fingerprints and evaluate Multiple Discriminant Analysis, Maximum Likelihood (MDA/ML) classification performance.

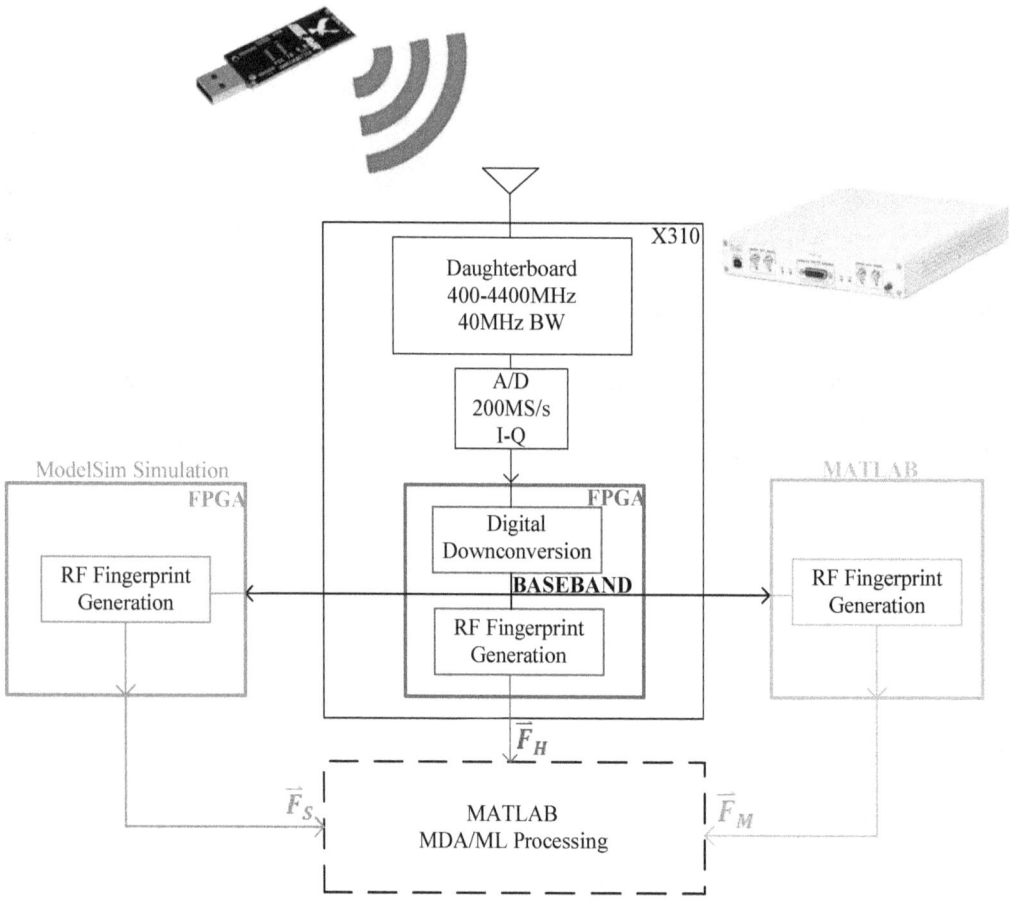

Figure 3.1: X310 SDR Methodology for Assessing RF-DNA Fingerprinting Using Matlab ($\vec{F}_M$), FPGA-Simulation ($\vec{F}_S$), and FPGA-Hardware ($\vec{F}_H$) Generated Fingerprints [30].

A modular FPGA design was planned and implemented using ModelSim FPGA simulation tools. The simulation model was validated using actual ZigBee bursts collected by the X310 SDR. The use of experimentally collected signals ensured realistic operation would be recreated as closely as possible. The FPGA simulation-generated fingerprints $\vec{F}_S$ were exported to MATLAB for MDA/ML classification evaluation. The FPGA design was compiled for use on the X310 SDR Kintex-7 FPGA and uploaded to the device. The X310 SDR collected emissions from multiple Zigbee devices and FPGA-Hardware generated $\vec{F}_H$ fingerprints using the embedded FPGA. Hardware generated $\vec{F}_H$ fingerprints were transferred to MATLAB and validated using both MDA/ML and Generalized Relevance Learning Vector Quantization-Improved (GRLVQI) classifiers.

Topics in this chapter are presented sequentially relative to the experimental methodology overview illustrated in Figure 3.1. Section 3.1 describes the X310 SDR configuration and setup procedures followed for the collection of radiated bursts. Section 3.2 describes the MATLAB model for FPGA fingerprint generation. Section 3.3 describes the FPGA fingerprinting implementation. Section 3.4 describes Dimensional Reduction Analysis (DRA) techniques used in this research. Section 3.5 details the process of device discrimination using the MDA/ML and GRLVQI classifiers.

## 3.1 Experimental Signal Collection

### 3.1.1 X310 SDR Receiver Configuration

The receiver used in this research was a National Instruments (NI) Universal Software Radio Peripheral (USRP) X310 Software-Defined Radio (SDR). The X310 SDR is a commercially available, inexpensive (approximately $5,000) SDR with transmit and receive capabilities covering DC to 6.0 GHZ depending on daughterboard installed. For this research, the SBX-40 daughterboard was installed and provided a receive frequency range of 400-4400 MHz with a maximum instantaneous bandwidth of $BW = 40.0$ MHz. A block diagram of the X310 SDR receiver architecture is shown in Figure 3.2.

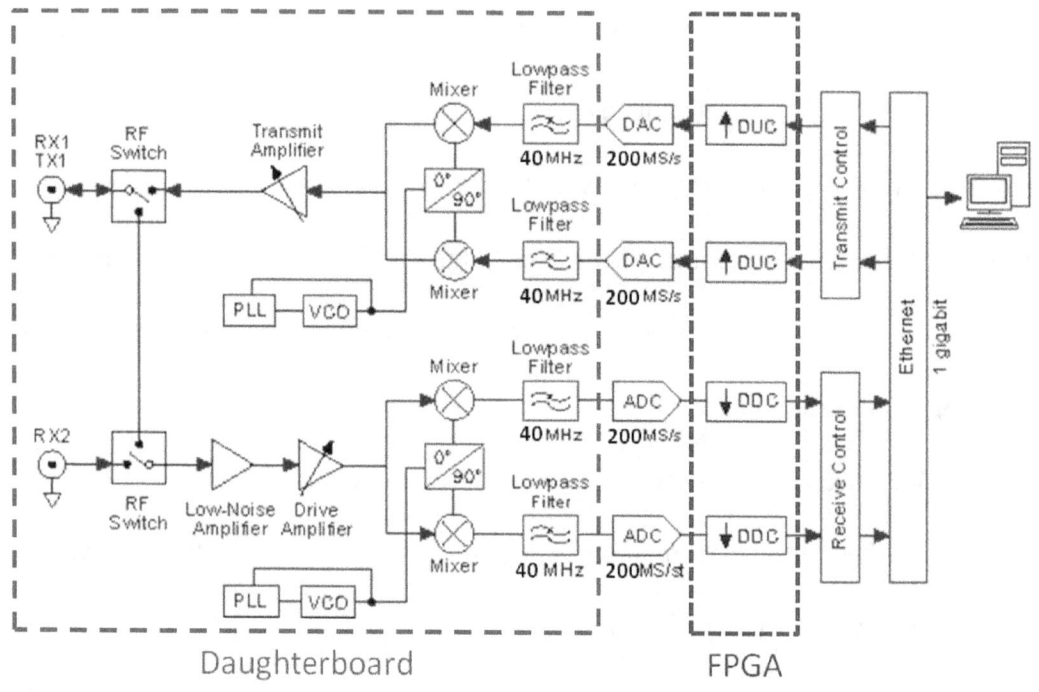

Figure 3.2: X310 SDR Receiver Architecture [6].

The RF emitting devices used in this research included $N_d = 5$ AVR RZUSBsticks. RZUSBstick is a device designed by Atmel Corporation for the development, debugging and demonstration of IEEE 802.15.4, 6LoWPAN, and ZigBee [1]. The RZUSBstick uses the Universal Serial Bus (USB) for configuration, transmission and reception of ZigBee data.

The RZUSBstick devices were connected to a computer running the open-source tool zbstumbler. Zbstumbler is an application from the killerbee suite, a popular collection of software tools used to manipulate ZigBee devices [21]. Zbstumbler was used to configure the devices to broadcast a ZigBee beacon request at a fixed interval indefinitely. The ZigBee channel used in this research was number $N_{ZC} = 26$ having a center frequency of $f_c = 2.480$ GHz. Channel $N_{ZC} = 26$ was chosen for consistency with

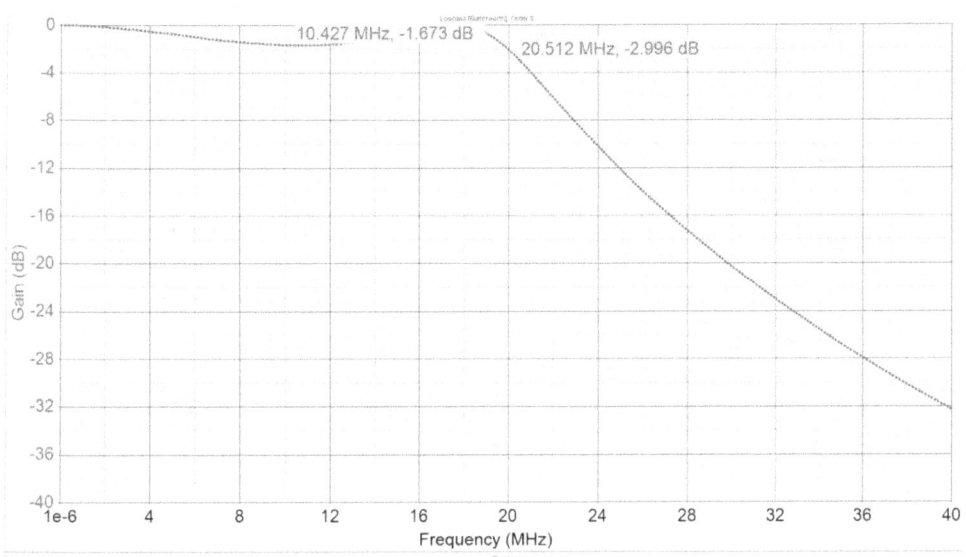

Figure 3.3: Normalized Frequency Domain Response of the SBX-40 Daughterboard Anti-Aliasing Filter (Positive Frequencies Only).

prior related work [20] to help mitigate interference from IEEE 802.11 WiFi. The SBX-40 instantaneous bandwidth is $BW = 40.0\,\mathrm{MHz}$ which was sufficient for collecting $BW_{RF} = 2.0\,\mathrm{MHz}$ ZigBee emissions as described in Section 2.1. The one-sided frequency domain response of the SBX-40 anti-aliasing filter is shown in Figure 3.3.

As illustrated in Figure 3.3, the bandwidth of the SBX-40 analog anti-aliasing filter is $BW_{0-3dB} = 20.5\,\mathrm{MHz}$. The collected ZigBee emissions were down-converted to an Intermediate Frequency (IF) using the analog mixer/anti-aliasing filter on the SBX-40. The complex waveform was sampled at $f_s = 200$ MS/s by the 14-bit dual channel Analog to Digital Converter (A/D) on the X310 SDR. The digital complex waveform was then sent to the FPGA to be digitally down-converted to baseband. The Digital Down-Conversion (DDC) chain in the FPGA is shown in Figure 3.4.

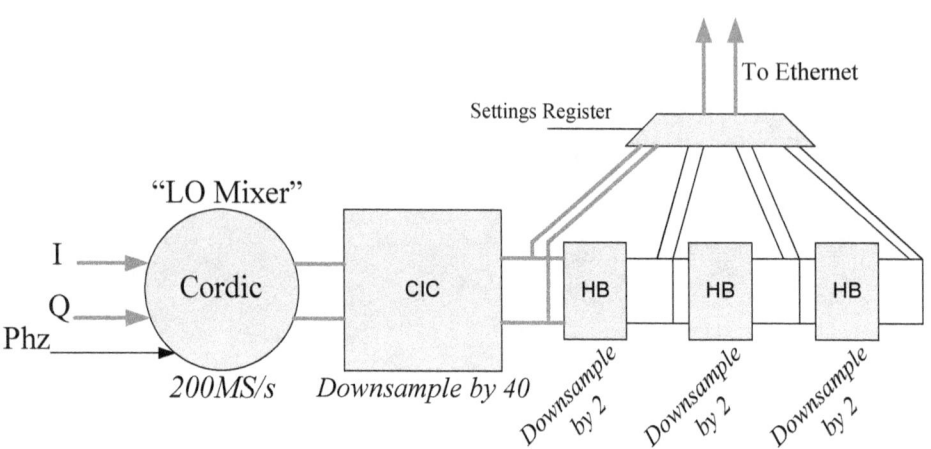

Figure 3.4: Digital Down-Conversion (DDC) Chain of X310 SDR with Received Signal Path in Red.

The DDC chain implemented uses a Coordinate Rotation Digital Computer (CORDIC) based mixer to down-convert the digitized signal to baseband. The CORDIC algorithm was implemented in rotation mode as described in Chapter 2. A Cascaded Integrator-Comb filter (CIC) was used to perform low-pass filtering and downsampling on the complex digital waveform. Unlike other common filters, the CIC has a decimator built into its architecture, simplifying the downsampling process. The CIC was chosen for this hardware application given its computational simplicity which only requires addition and subtraction operations for the CIC filter design. The CIC filter block diagram is shown in Figure 3.5.

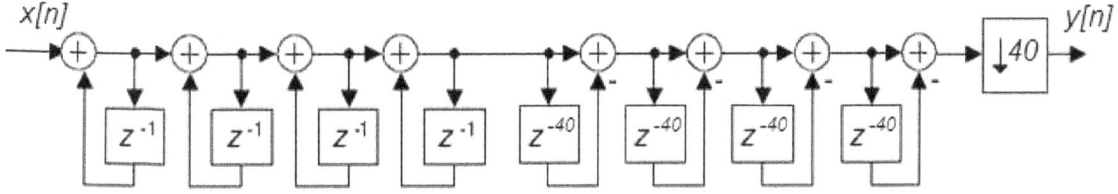

Figure 3.5: 4-Stage $r_d = 40$ CIC Filter [15].

The CIC design consists of $n_s = 4$ "integrator" addition stages followed by $n_s = 4$ "comb" subtraction stages and an $r_d = 40$ decimator. The CIC was implemented as described in [8]. The normalized CIC frequency response is shown in Figure 3.6.

The CIC decimates the $f_s = 200$ MS/s signal to a new sample rate of $f_s = 5$ MS/s. The sample rate of $f_s = 5$ MS/s is suitable for the previously described $BW_{RF} = 2.0$ MHz bandwidth of ZigBee channel $N_{ZC} = 26$. The sample rate of $f_s = 5$ MS/s was experimentally deemed sufficient for correct demodulation of ZigBee emissions. The CIC also performs filtering on the complex waveform. The digital down-conversion chain implemented also contains selectable half-band filters for use with different sample rates, but they are not selected in this research. The baseband signal is then routed out of the X310 SDR FPGA and to the computer via Ethernet.

### 3.1.2 X310 SDR Receiver Configuration

All collections were made within a shielded Ramsey STE6000 test enclosure to ensure a low-noise collection environment for initial proof-of-concept demonstration. The commercial shielded test enclosure used is shown in Figure 3.7.

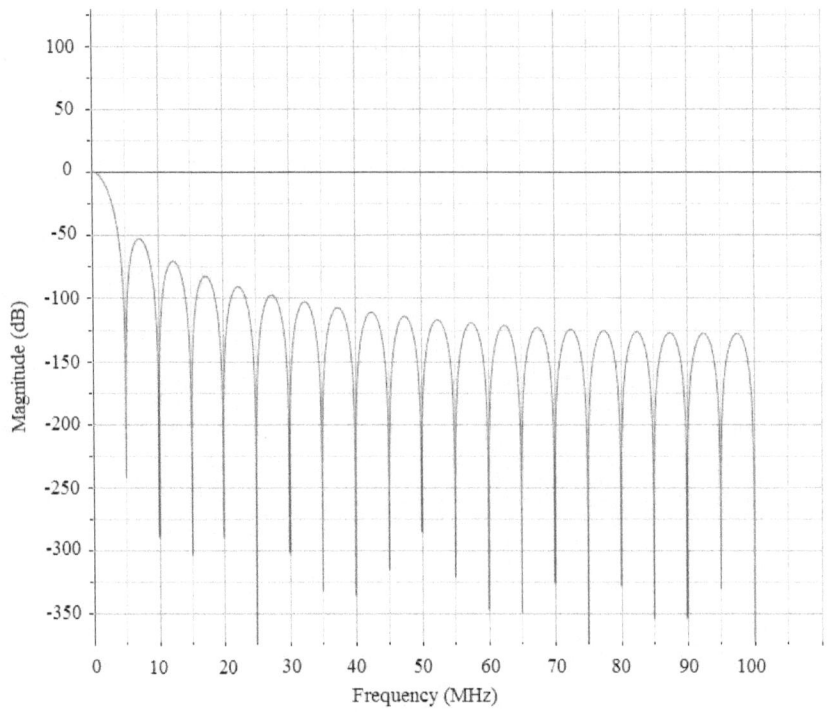

Figure 3.6: Normalized Frequency Response of 4-stage $r_d = 40$ CIC Filter.

Figure 3.7: Ramsey STE6000 Shielded Test Enclosure [19].

The Ramsey STE6000 was chosen because it was designed for use with Industrial, Scientific and Medical (ISM) band signals such as Bluetooth, WiFi and ZigBee [19]. The STE6000 provides over 90.0 dB of attenuation at $f = 2.0$ GHz according to manufacturer specification. Additionally, the interior has an RF absorbent foam coating that attenuates by $a = 24.0$ dB, mitigating multipath interference. The STE6000 was equipped with Ethernet and USB connections so it can remain closed while controlling the X310 SDR and ZigBee device, respectively.

The ZigBee Device Under Test (DUT) was connected to the internal USB port of the STE6000. The X310 SDR was positioned such that its antenna was $l = 8.0$ cm from the DUT. The STE6000 was closed and sealed. The computer was used to configure the X310 SDR FPGA with the proper firmware. For signal collection, the receiver mode firmware was flashed onto the FPGA. For RF-DNA fingerprint generation, the fingerprint generation mode firmware was flashed onto the FPGA. Zbstumbler software was initiated to configure the DUT to broadcast ZigBee beacon requests at ZigBee channel $N_{ZC} = 26$ with a rate of $R_b = 10$ $b/s$ (bursts per second). The X310 SDR was then configured to initiate the start of signal collection. Received signals (or generated fingerprints) were then streamed over Ethernet to the computer where they were saved to the hard-disk. This process was repeated for all $N_d = 5$ devices.

## 3.2 MATLAB Model for FPGA Fingerprint Generation

To demonstrate fingerprint generation in real-time on the FPGA platform, it was desirable to reduce the computational complexity of previous fingerprint generation methods. FPGA operations occur in real-time and on synthesized hardware. Certain operations that can be easily calculated with enough time in MATLAB are costly in fixed hardware resources on an FPGA. For this reason, a reduced-complexity fingerprinting model was developed with MATLAB as a reduced-complexity subset of the traditional fingerprinting process.

### 3.2.1 Burst Detection

Accurate burst detection of the target signal is critical to the RF-DNA process. If the collected bursts are not properly aligned, the statistical features will cross the region boundaries, which will degrade process performance process. Therefore, the first step in the RF-DNA process is the alignment of collected bursts. A burst detection algorithm was developed that could operate in real-time, with low latency and with low computational complexity. A squaring-smoothing amplitude detection algorithm was applied to the real-valued waveform *Re* of the incoming signal as follows:

$$g[n] = \frac{(Re[n])^2 + (Re[n+1])^2 + \cdots + (Re[n+31])^2}{32}. \tag{3.1}$$

The resultant value $g$ is compared to a threshold value $k$. When $g[n] > k$, the start of the burst is indicated as $g[n-30]$ to collect the transient turn-on region of the transmitted waveform. Though the *Re* component was used to detect the burst, similar results can be obtained using the signal *Im* component. A representative output of the smoothing algorithm is shown in Figure 3.8.

The threshold value $k$ was empirically chosen to give consistent burst detection performance. Knowing the duration of the ZigBee preamble and the starting sample number, we can select the beginning and end of the ZigBee preamble waveform. The detected preamble was then extracted from the signal for fingerprint generation.

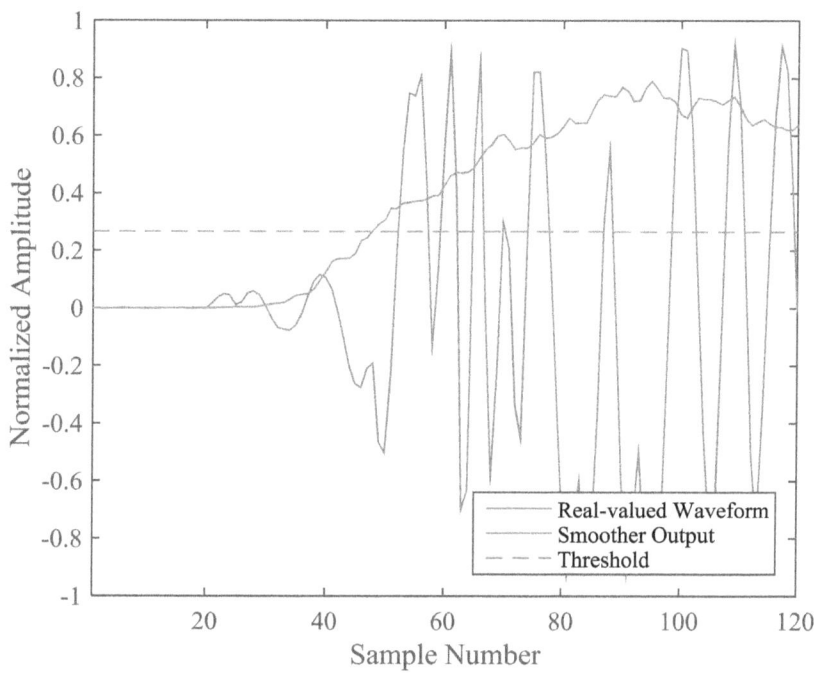

Figure 3.8: Output of Squaring-Smoothing Burst Detection Algorithm.

### 3.2.2 Fingerprint Generation

The ZigBee waveform preamble is the region of interest (ROI) for ZigBee fingerprint generation. The preamble duration is standardized to be $t_p = 128.0$ μs per the IEEE 802.15.4 ZigBee specification [9]. At a sample rate of $f_s = 5$ MS/s this corresponds to a length of $N_s = 640$ samples. The ROI is further separated into $N_R = 10$ equal length subregions. Because fingerprints are calculated in real-time on the FPGA, odd-numbered subregions ([1, 3, 5, 7, 9]) are received and processed during the time of even numbered subregions ([2, 4, 6, 8, 10]). Therefore, the even numbered subregions are excluded from the product fingerprint. This process yields a fingerprint comprised of features based on $N_R = 5$ subregions. Figure 3.9 shows a ZigBee amplitude response collected by the X310 SDR with $N_R = 10$ subregions highlighted.

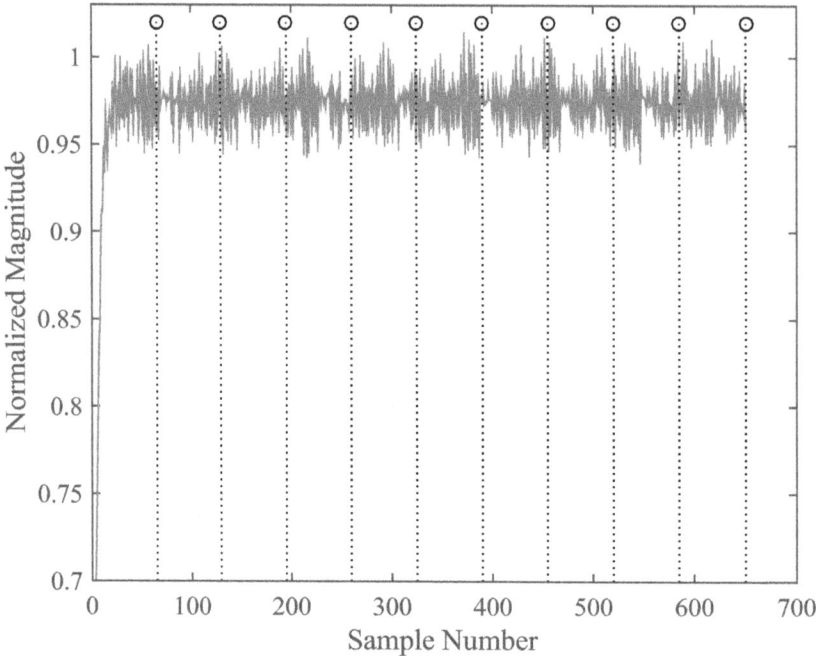

Figure 3.9: Experimentally Collected ZigBee Preamble with $N_R = 10$ Subregions.

The collected amplitude waveform shown in Figure 3.9 is $N_s = 650$ samples (130 μs) in length, which is approximately equal to the standard length of $t_p = 128.0$ μs from the IEEE specification. In the simplified MATLAB model, the following time-domain responses were chosen for fingerprint generation:

1. Real-valued ($Re[n]$) time domain waveform,

2. Imaginary-valued ($Im[n]$) time domain waveform,

3. Instantaneous phase ($\phi[n]$) given by

$$\phi[n] = tan^{-1}\left(\frac{Im[n]}{Re[n]}\right), \tag{3.2}$$

4. Instantaneous amplitude ($a[n]$) given by

$$a[n] = \sqrt{(Re[n])^2 + (Im[n])^2}. \tag{3.3}$$

For each of $N_R = 5$ subregions of the $Re[n]$, $Im[n]$, $\phi[n]$ and $a[n]$ waveforms, the variance ($\sigma^2$) of each respective waveform was calculated and concatenated to form fingerprint vector $\vec{F}_M$ as follows:

$$\vec{F}_M = [\sigma^2_{\phi_{R1}}, \sigma^2_{a_{R1}}, \sigma^2_{Re_{R1}}, \sigma^2_{Im_{R1}} \cdots \sigma^2_{\phi_{R5}}, \sigma^2_{a_{R5}}, \sigma^2_{Re_{R5}}, \sigma^2_{Im_{R5}}]. \tag{3.4}$$

Fingerprint vector $\vec{F}_M$ was then rounded to 32-bit fixed-point decimal to match the 32-bit output capability of the X310 SDR. A total of $N_b = 1000$ bursts were transmitted by each of $N_d = 3$ devices and received by the X310 SDR at a sample rate of $f_s = 5$ MS/s. A fingerprint vector $\vec{F}_M$ was generated for each burst received, and an MDA/ML classification model was built and evaluated as described in Chapter 2.

## 3.3 FPGA Fingerprint Generation

The X310 SDR has a Xilinx Kintex-7 FPGA which contained the DDC chain as described in Section 3.1. A fingerprint generator was designed for the Kintex-7 with the purpose of implementation on the X310 SDR FPGA. The fingerprint generator design was then simulated with ModelSim FPGA simulation software to generate the FPGA-Simulation fingerprint vector $\vec{F}_S$. Generation of FPGA-hardware fingerprint vector $\vec{F}_H$ was then accomplished by synthesizing the fingerprint generator design within the X310 SDR FPGA.

### 3.3.1 FPGA Fingerprint Generator Design

The block diagram for the developed FPGA fingerprint generator design is shown in Figure 3.10.

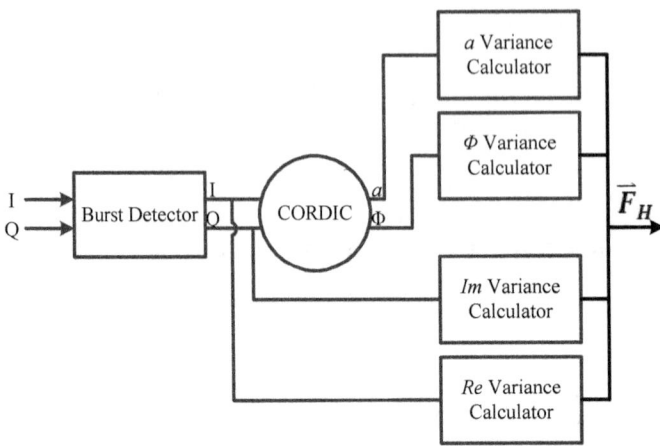

Figure 3.10: Fingerprint Generator Design as Implemented on X310 SDR FPGA.

The design was developed to match MATLAB model functionality as described in Section 3.2. A CORDIC module was implemented in vectoring mode as described in Chapter 2. The CORDIC module was used to calculate the instantaneous phase $\phi[n]$ and instantaneous amplitude $a[n]$ from $Re[n]$ and $Im[n]$. The functional component groups of the FPGA fingerprinting design are:

1. A squaring-smoothing amplitude-based burst detector as described in Section 3.2.1.

2. A CORDIC module operating in vectoring mode as described in Chapter 2.

3. Variance calculating modules for each of $N_f = 4$ instantaneous feature waveforms. Variance calculators were implemented in parallel. Because of parallel implementation, all $N_f = 4$ variance values for a given subregion are calculated simultaneously.

After calculation of $N_f = 4$ variance values for the given subregion, results were sent out of the X310 SDR for concatenation into the full-dimensional fingerprint vector.

### 3.3.2 Simulated FPGA Fingerprint Generation

To characterize any coloration effects inherent to the FPGA-implemented fingerprint generator design, a simulation model was desired. The fingerprint design as described in Section 3.3.1 was implemented in a ModelSim FPGA simulation environment. Experimentally collected ZigBee bursts were used as input to the ModelSim simulated design, and the simulation-generated fingerprints $\vec{F}_S$ were collected and stored for classification performance analysis. This simulation setup is shown in Figure 3.11.

To build a classification model as described in Chapter 2, a total of $N_b = 1000$ bursts were simulated as collected from each of $N_d = 3$. Simulation-generated fingerprints $\vec{F}_S$ were collected and used for model development results as presented in Chapter 4.

### 3.3.3 Hardware FPGA Fingerprint Generation

To characterize the real-time performance of the FPGA-based fingerprint generator, the fingerprint generator module was inserted at the end of the X310 SDR DDC chain and instantiated in the actual FPGA hardware. The block diagram for the resultant X310 SDR hardware chain is shown in Figure 3.12.

As shown in Figure 3.12, the FPGA-based hardware implementation allows for hardware-generated fingerprints $\vec{F}_H$ to be streamed to the computer as they are received by the X310 SDR. A total of $N_b = 1000$ bursts were collected and processed from each

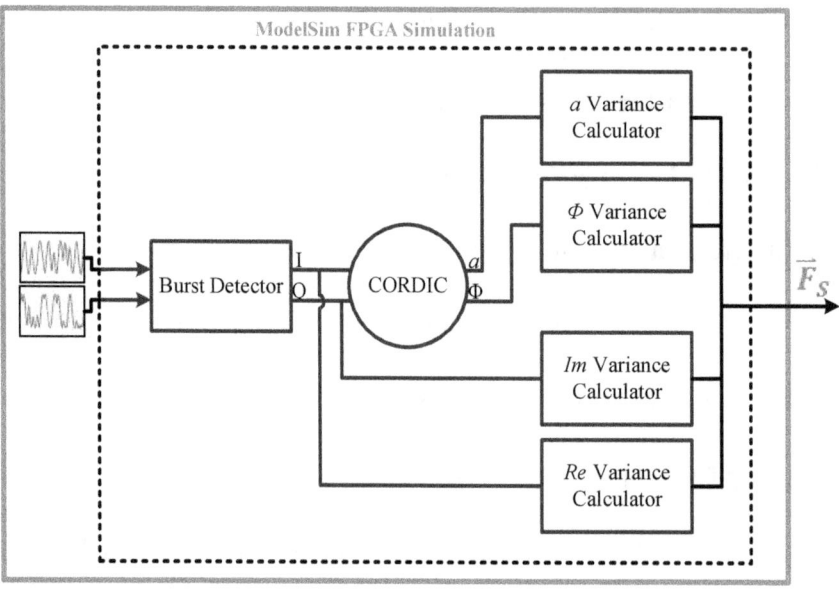

Figure 3.11: ModelSim FPGA Simulation Environment for Generation of $\vec{F}_S$ from X310 SDR Collected ZigBee Bursts.

of $N_d = 3$ to generate FPGA-Hardware fingerprints $\vec{F}_H$. The resultant fingerprints were used to build a model and evaluate classification performance as presented in Chapter 4.

**3.4 Feature Set Dimensional Reduction**

The complete RF-DNA fingerprint used in this research is based on $N_R = 5$ subregions with $N_f = 4$ instantaneous feature waveforms. The length $N_D$ of a full-dimensional fingerprint is:

$$N_D = N_R \times N_f \qquad (3.5)$$

Therefore, a full-dimensional fingerprint of length $N_D = 20$ was used in this research. A process known as Dimensional Reduction Analysis (DRA) was performed to limit the number of features used for model development. The purpose of DRA is to determine which features can be eliminated while maintaining the desired classification

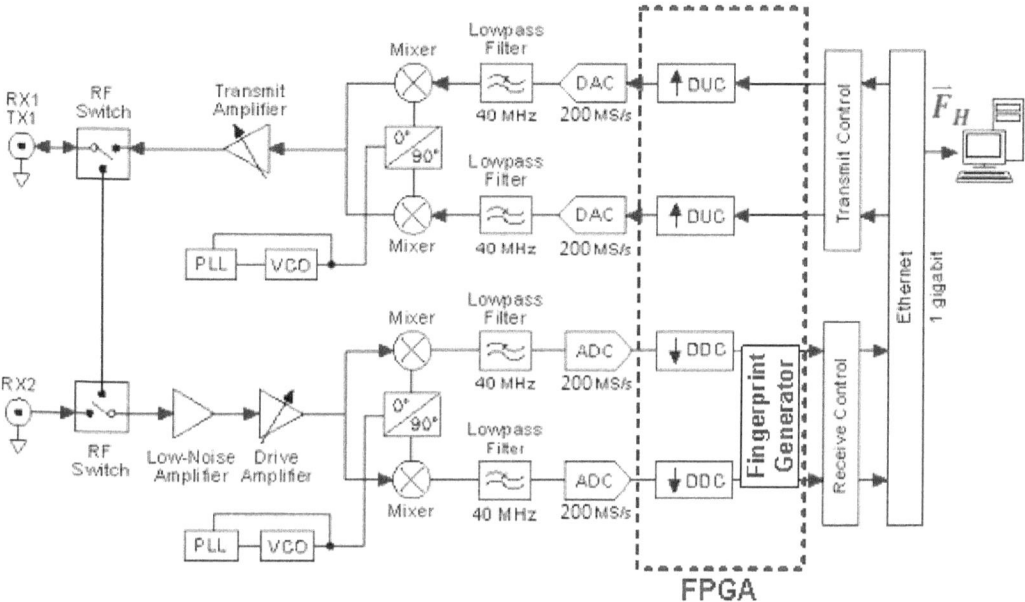

Figure 3.12: X310 SDR Hardware Chain with FPGA-based Fingerprint Generator [6].

performance level. This research compared two methods of DRA: *qualitative* and *quantitative*.

*Qualitative* DRA was performed by selecting all features from a particular feature subset: *a*-only, $\phi$-only, *Re*-only or *Im*-only. Each dimensionally reduced feature subset was used to form $N_D = 5$ length fingerprints based only on that subset.

*Quantivative* DRA was performed by selecting only the top-5 most relevant features as determined by the GRLVQI process. All DRA feature subsets are displayed in Table 3.1.

Table 3.1: Dimensionally Reduced Feature Sets used for MDA/ML Classification

| DRA Method | Feature Set | $N_D$ |
|---|---|---|
| Full Dimensional | All | 20 |
| Qualitative | *a*-only | 5 |
| Qualitative | $\phi$-only | 5 |
| Qualitative | *Re*-only | 5 |

| Qualitative | *Im*-only | 5 |
| Quantative | GRLVQI Top-5 | 5 |

## 3.5 Device Classification

After $\vec{F}_M$, $\vec{F}_S$ and $\vec{F}_H$ fingerprints were generated from $N_b = 1000$ bursts each, and for each of $N_D = 3$ devices, they were input to the MDA/ML or GRLVQI device discrimination process. MDA/ML and GRLVQI were performed as described in Chapter 2. The fingerprints were first separated into equal length *training* and *testing* fingerprint sets. *Training* and *testing* fingerprint sets were taken as interleaved (odd and even) subsets of the complete fingerprint set. The *training* set was then input to the MDA/ML and GRLVQI classifiers where the classification models were developed. A $K$-fold cross-validation process was used by both classifiers to determine the "best" model using a $K = 5$ value. Once models were developed by both classifiers, the *testing* set of fingerprints was used to assess classification performance. Fingerprints were then classified in a "Looks most like?" assessment, assigning each *testing* fingerprint to the device it was estimated to be. The above steps were repeated for dimensionally reduced fingerprint vectors. These results are presented in Chapter 4.

## 3.6 Device ID Verification

While device classification performed a "Looks most like?" comparison, device ID verification provides a "Looks how much like?" assessment. In the target air monitor application, verification will be used to reject "rogue" devices that are not authorized network access.   This rogue rejection will enhance network security and augment traditional bit-level security techniques.

Device ID verification was implemented as described in Chapter 2.   After model development, each *testing* fingerprint was compared to each of $N_D = 3$ devices on which the model was based.  The verification test statistics $z_v$ provides a measure of similarity between the compared pair of devices.   The test statistic $z_v$ is then compared to a verification threshold value $t_v$ to make a binary decision of accepting or rejecting the device's claimed identity.   By comparing only authorized devices to one another and varying the threshold value $t_v$, the relationship between True Verification Rate (TVR) and False Verification Rate (FVR) was explored.   TVR is the percentage of instances where a device is correctly authorized after claiming its own identity.   FVR is the percentage of instances where a device is authorized after claiming an identity that is not its own.

An additional case was explored where "rogue" devices were introduced to the system claiming the identity of authorized devices.   The relationship between TVR and Rogue Accept Rate (RAR) was determined.   RAR is the percentage of instances where a rogue device is incorrectly accepted as an authorized device.   Results for TVR vs. FVR and TVR vs. RAR are presented in Chapter 4.

# IV. Results and Analysis

## 4.1 Introduction

This chapter provides results for Radio Frequency Distinct Native Attribute (RF-DNA) discrimination of ZigBee devices, to include comparison of fingerprinting performance using MATLAB-generated fingerprints, Field Programmable Gate Array (FPGA)-generated fingerprints, and fingerprints generated in a simulated FPGA environment. Additional results are analyzed, including authorized and rogue device verification using a Multiple Discriminant Analysis, Maximum Likelihood (MDA/ML) mode. Classification performance is also analyzed using reduced dimensional feature sets (proper subsets of full dimensional feature sets) as well as FPGA timing and utilization results.

This chapter is organized as follows: Section 4.2 presents Time-Frequency (T-F) analysis of experimentally collected ZigBee signals that were collected using the X310 Software-Defined Radio (SDR). Section 4.3 presents classification results of the MDA/ML model comparing three fingerprint generation methods: 1) simulated-based MATLAB generated fingerprints ($\vec{F}_M$), 2) simulation-based FPGA generated fingerprints ($\vec{F}_S$), and 3) hardware-based X310 FPGA-generated fingerprints ($\vec{F}_H$). Section 4.4 presents authorized and rogue device model verification results. Section 4.5 presents classification performance results using both *qualitative* Dimensional Reduction Analysis (DRA) as well as the *quantitative* DRA based on feature relevance ranking using a Generalized Relevance Learning Vector Quantization-Improved (GRLVQI) process.

Finally, Section 4.6 provides timing results for the FPGA hardware design as well as device capacity utilization.

## 4.2 Time Frequency (T-F) Analysis

A Time-Frequency (T-F) analysis was conducted to investigate the spectral and temporal characteristics of experimentally collected ZigBee signals that were collected using the X310 SDR. Two types of T-F analysis were performed to validate the experimental collection setup, including 1) analysis of the X310 SDR internal background noise and 2) analysis of experimentally collected ZigBee emission characteristics and comparison with standard specifications.

### 4.2.1 X310 Background Noise Analysis

The X310 SDR receiver was operated inside a shielded test enclosure without any other devices present. These collections were used to characterize internal noise to the X310 SDR. Additionally, this analysis showed the effectiveness of the shielded test enclosure in attenuating outside radiation. The noise environment was sampled at a sample rate of $f_s = 5$ MS/s and center frequency of $f_c = 2.480$ GHz. These collection parameters remained constant for all collections and analysis conducted the research. The resultant normalized Power Spectral Density (PSD) for a one minute X310 background noise collection is shown in Figure 4.1.

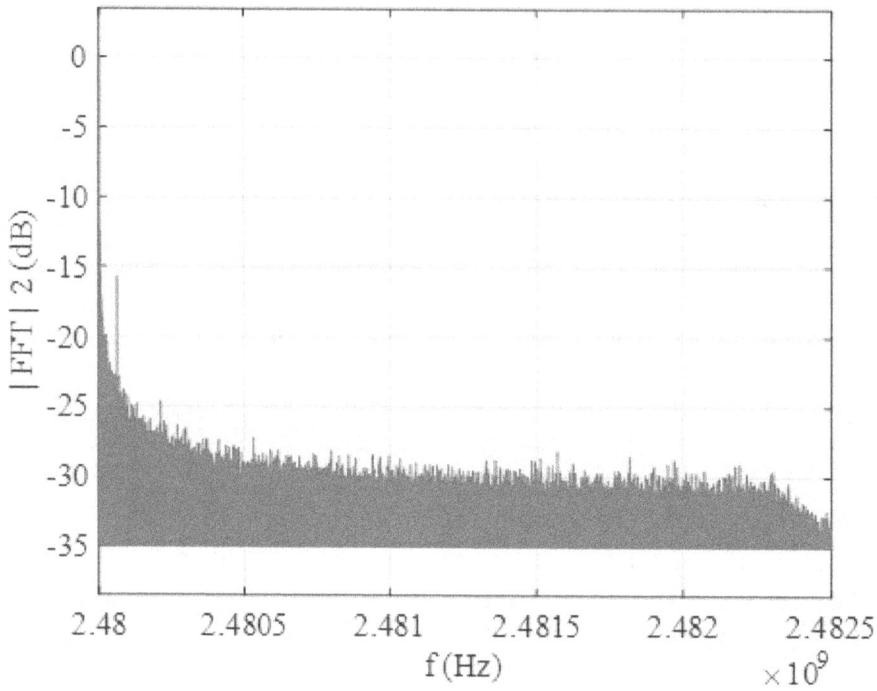

Figure 4.1: Normalized X310 SDR Background Noise PSD.

As shown in Figure 4.1, the noise environment has a small peak at $f = 2.4861$ GHz. The results obtained in this research are not affected by the internal noise of the receiver because it is present equally for all devices.

### 4.2.2 Collected ZigBee Emission Analysis

Using the collection methodology described in Section 3.3, ZigBee beacon requests (bursts) were collected with the X310 SDR. The time domain response of an experimentally collected Zigbee burst is illustrated in Figure 4.2:

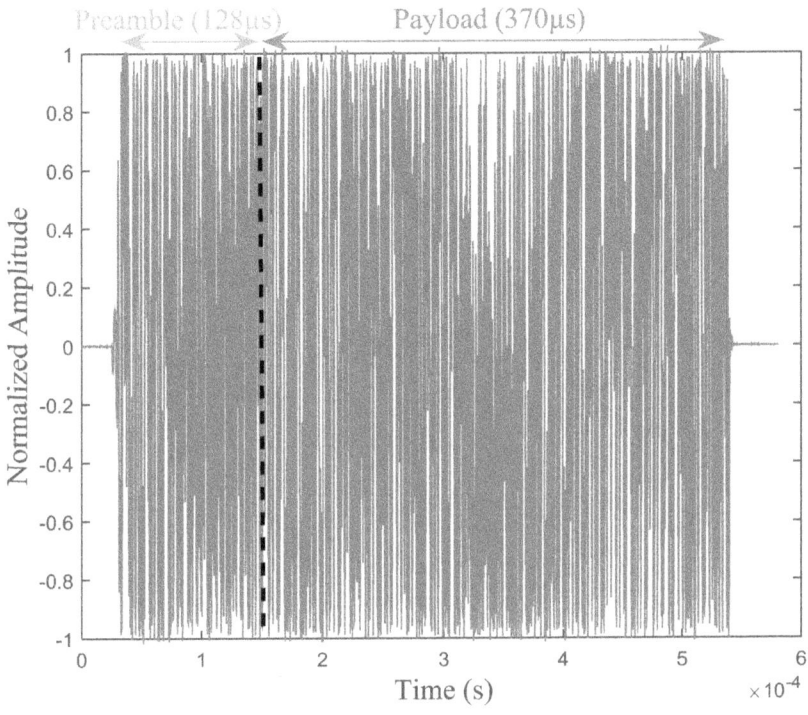

Figure 4.2: Representative X310 SDR Collected ZigBee Burst Amplitude Response Showing Preamble and Payload Regions.

As described in Section 2.1, the ZigBee preamble consists of 8 Offset Quadrature Phase Shift Keying (O-QPSK) modulated symbols. Figure 4.3 shows the preamble response of a typical ZigBee burst, divided into eight symbols. The emissions were collected near-baseband to enhance the visibility of the information in the figure.

ZigBee device transmissions were initiated by controlling the device with a computer running the Zbstumbler script. The zbstumbler script is an open source application from the killerbee suite [7], a popular collection of software tools used to manipulate ZigBee devices. The ZigBee devices were configured to broadcast at a transmission rate of $R_b = 10 \ b/s$. The frequency domain response for a $t = 0.55 \ s$ collection is shown in Figure 4.4.

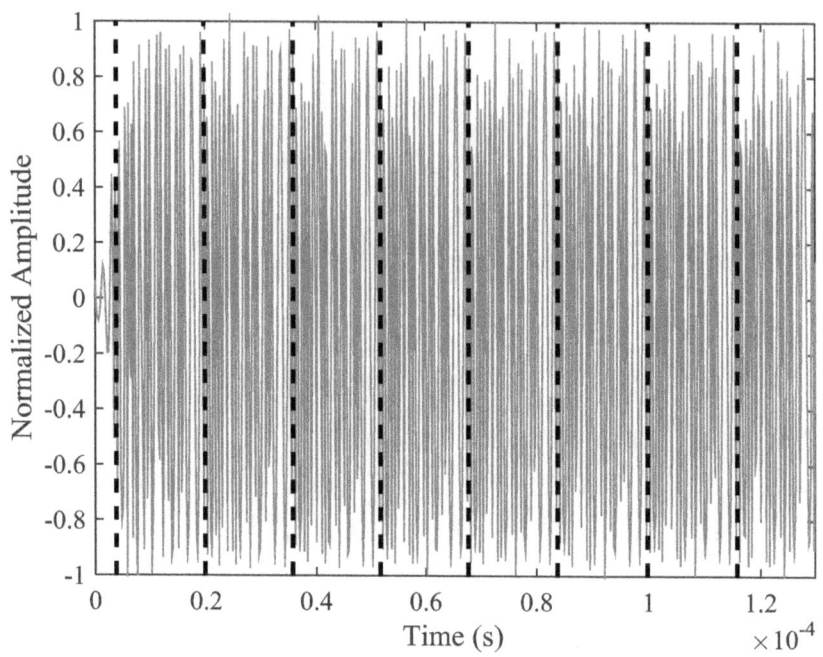

Figure 4.3: ZigBee Preamble Response from Fig. 4.2 Showing 8 O-QPSK Symbols.

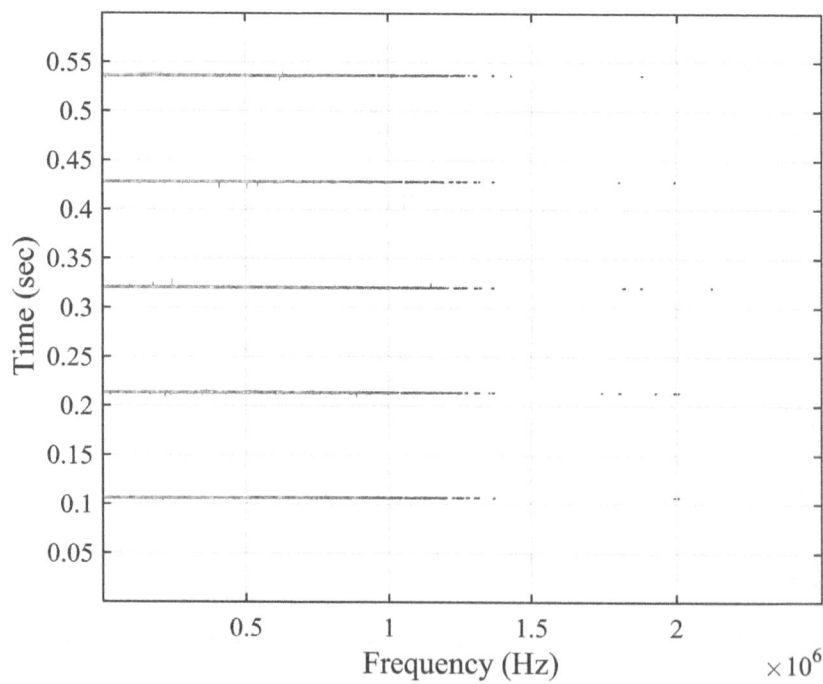

Figure 4.4: Frequency Response Over Time of $N_b = 5$ Baseband ZigBee Bursts.

Although the burst rate was configured to be $R_b = 10\ b/s$, the actual rate of burst transmission was $R_b \approx 9\ b/s$. A total of $N_b = 20$ bursts were collected at sample rate $f_s = 5\ MS/s$ and down-converted to baseband. The $N_b = 20$ bursts were then overlaid and averaged. The normalized Power Spectral Density (PSD) of the resultant waveform is shown in Figure 4.5.

As described in Section 2.1, the specified bandwidth of a ZigBee channel is $BW_{RF} = 2.0\ MHz$. Therefore, the single-sided baseband bandwidth of a ZigBee channel is $BW_{BB} = 1.0\ MHz$. There is a 7 dB decrease in power at $BW_{BB} = 1.0\ MHz$. These results are illustrated in Figure 4.5.

**4.3 Classification Model Development**

Using the X310 SDR operating in fingerprint generation mode inside a shielded test enclosure, a total of $N_b = 1000$ ZigBee bursts were collected per device. The *collected* Signal-to-Noise Ratio was on the order of $SNR_c \approx 55.0\ dB$ for all of the $N_d = 5$ devices. A total of $N_p = 1000$ fingerprints were generated by the X310 SDR. Using MDA, a classification model was developed based on the fingerprints of Atmel RZUSBstick devices 1, 2 and 3. The model was developed using $N_{Tng} = 500$ *training* fingerprints per device, extracted from a larger pool of $N_T = 1000$ total fingerprints, with full-dimensional fingerprints generated on the FPGA.

A $K$-fold cross-validation process was used to determine the "best" MDA model using a $K = 5$ value. A model was developed that maximized Euclidian distance between device/class means. That model was used as the projection matrix $\boldsymbol{W}_B$.

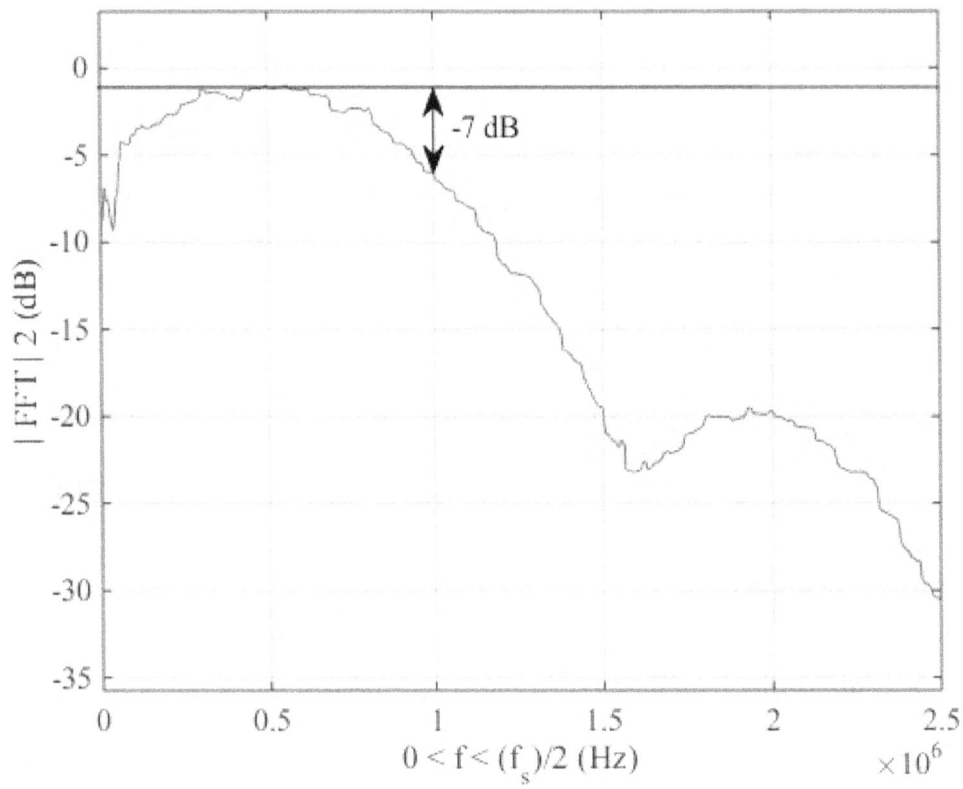

Figure 4.5: Normalized PSD of $N_b = 20$ Averaged Baseband ZigBee Bursts.

Projection matrix $\boldsymbol{W_B}$ was then multiplied by each of $N_{Tst} = 500$ *testing* fingerprints per device to project the fingerprint into a 2-dimensional Fisher space. The model accuracy was quantified using average percent correct classification (*%C*) based on testing fingerprint classification performance for each device, as well as a cross-class *%C* for all devices. These results are shown in Figure 4.6.

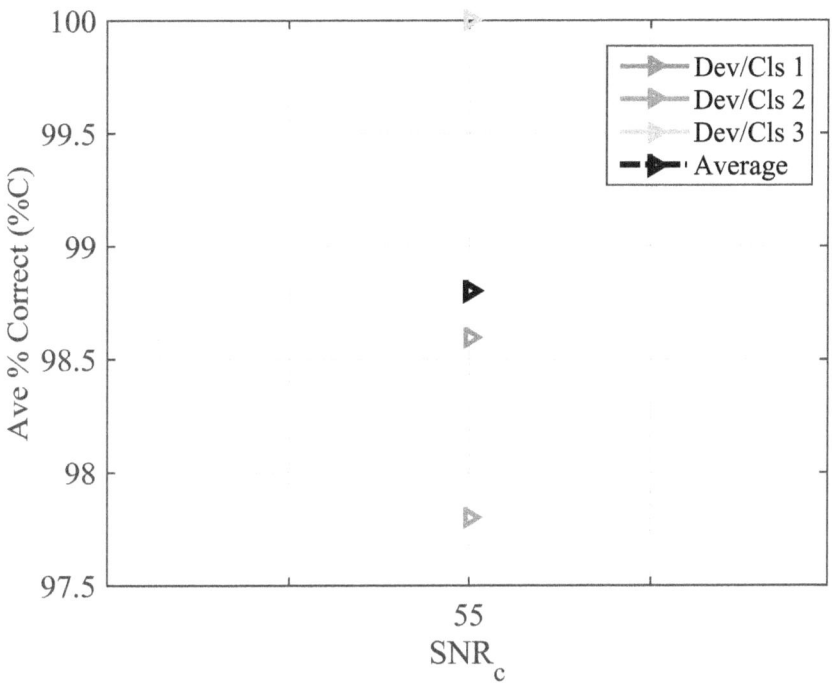

Figure 4.6: MDA/ML %$C$ for each Class/Device and the Cross-Class/Device Average %$C$ at $SNR_c \approx 55.0$ dB using $N_{Tst} = 500$ Fingerprints per Class/Device.

In addition to cross-class %$C$ as shown in Figure 4.6, the tendency for devices to be incorrectly classified as each other was also quantified and reflected in the classification confusion matrix as shown in Table 4.1.

As shown in Table 4.1, all devices achieved an arbitrary benchmark of correct classification rate %$C > 90\%$. Additionally, while device 3 achieved %$C = 100\%$, devices 1 and 2 "looked like" each other, resulting in some misclassification between the two devices. New models for $N_d = 4$ and $N_d = 5$ devices were created to assess the performance of the system with additional devices. Tables 4.2 and 4.3 show the confusion matrices for $N_d = 4$ and $N_d = 5$ class problems, respectively.

Table 4.1: Confusion Matrix of $N_d = 3$ Class Problem at $SNR_c \approx 55.0$ dB.

|  |  | Classified As | | |
|---|---|---|---|---|
|  |  | Dev 1 | Dev2 | Dev 3 |
| Input | Dev 1 | 98.6% | 1.4% | 0% |
|  | Dev 2 | 2.2% | 97.8% | 0% |
|  | Dev 3 | 0% | 0% | 100% |

Table 4.2: Confusion Matrix of $N_d = 4$ Class Problem at $SNR_c \approx 55.0$ dB.

|  |  | Classified As | | | |
|---|---|---|---|---|---|
|  |  | Dev 1 | Dev 2 | Dev3 | Dev 4 |
| Input | Dev 1 | 98.8% | 1.2% | 0% | 0% |
|  | Dev 2 | 1.8% | 98.2% | 0% | 0% |
|  | Dev 3 | 0% | 0.2% | 99.6% | 0.2% |
|  | Dev 4 | 0% | 0% | 0% | 100% |

Table 4.3: Confusion Matrix of $N_d = 5$ Class Problem at $SNR_c \approx 55.0$ dB.

|  |  | Classified As | | | | |
|---|---|---|---|---|---|---|
|  |  | Dev 1 | Dev 2 | Dev 3 | Dev 4 | Dev 5 |
| Input | Dev 1 | 98.6% | 1.4% | 0% | 0% | 0% |
|  | Dev 2 | 1.2% | 98.8% | 0% | 0% | 0% |
|  | Dev 3 | 0% | 0% | 99.8% | 0.2% | 0% |
|  | Dev 4 | 0% | 0% | 0.2% | 99.8% | 0% |
|  | Dev 5 | 0% | 0% | 0% | 0% | 100% |

## 4.4 Dimensional Reduction Analysis (DRA)

Because of the algorithm used by MDA/ML in generating a classification model, the specific features which give the best classification performance cannot be determined. Dimensional Reduction Analysis (DRA) techniques can be used to identify a selected subset of features that provides acceptable classification performance. Dimensional reduction can be achieved using qualitative methods or quantitative methods. Examples of qualitatively selected feature subsets include amplitude-only (Amp), phase-only (Phz), Real-only (Re) and Imaginary-only (Im). A GRLVQI classifier, described in Section 2.4, was also used to quantitatively select a subset of features based on their respective

influence on correct device classification. Figure 4.7 shows classification performance results of an initial trial comparing qualitatively selected features and the top-ranked $N_f = 5$ quantitatively selected features as selected by GRLVQI. The best performance of $\%C > 90\%$ is obtained by using the full-dimensional set of $N_f = 20$ features. Additionally, the FPGA hardware performance was contrasted with the performance of a simulated FPGA environment. These results are also shown in Figure 4.7.

Additional FPGA hardware trials were performed for a total of $N_{Trl} = 20$ trials. The performance results from $N_{Trl} = 20$ trials are shown with 95% confidence intervals (CI) in Figure 4.8. The mean $\%C$ for each DRA subset is also shown in Figure 4.8 with 95% confidence intervals omitted because they are within the vertical extent of the markers.

As shown in Figure 4.8, the full-dimensional $N_f = 20$ fingerprints consistently exceeded the arbitrary benchmark of $\%C = 90\%$ for all $N_{Trl} = 20$ trials. Cross-trial mean results for $N_f = 5$ Amp, Re, and Im are statistically equivalent based on 95% CI. Finally, performance was poorest for the $N_f = 5$ Phz DRA fingerprints which consistently yielded the lowest $\%C$ classification performance.

## 4.5 Device Verification and Rogue Detection

Device verification allows for a comparison between devices to describe "how much alike" the devices are. By making this comparison, a relationship can be found between the True Verification Rate (TVR) and False Verification Rate (FVR). TVR is the rate at which a device, claiming to be itself, is correctly authorized. FVR is the rate at which an unauthorized device, claiming to be the authorized device, is incorrectly

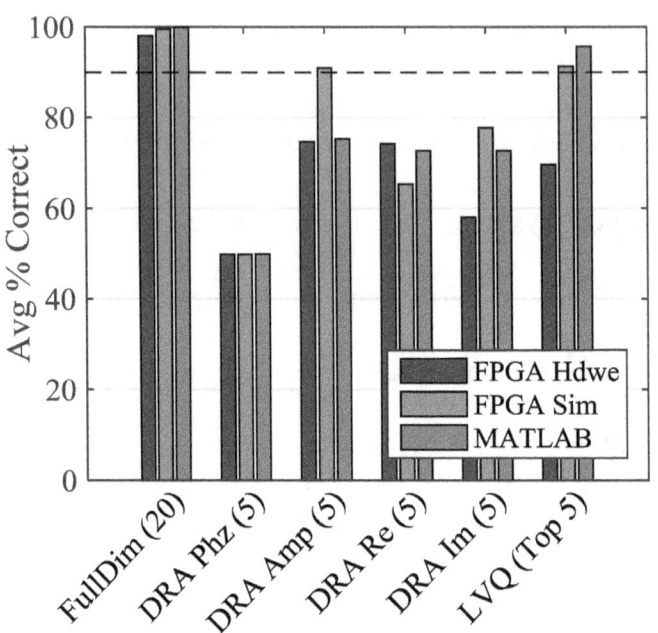

Figure 4.7: Comparison of Qualitative (Phz, Amp, Re, Im) and Quantitative (LVQ) DRA Performance with Full Dimensional Performance using $\vec{F}_M$, $\vec{F}_S$, and $\vec{F}_H$ Generated Fingerprints. The Number of Features per Feature Set is Indicated in Parenthesis.

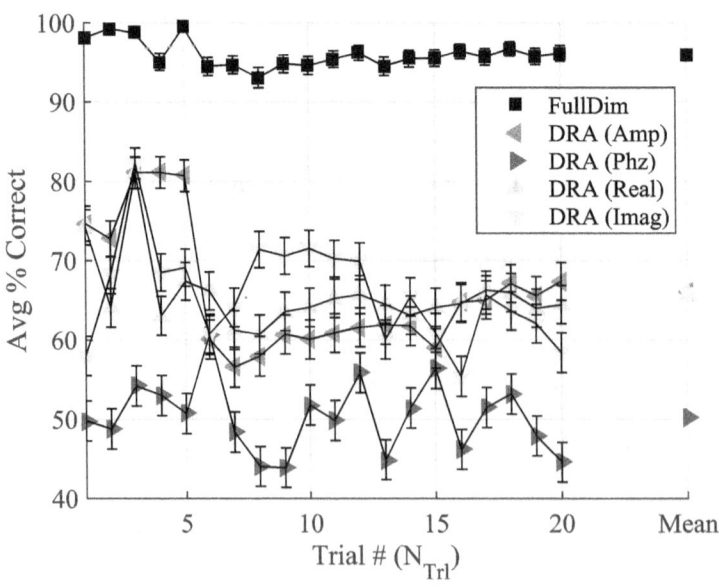

Figure 4.8: Average Percent Correct Classification (%C) with 95% Confidence Intervals for a Total of $N_{Trl} = 20$ Independent Experimental FPGA Hardware Trials. The Cross-Trial Mean Shows that Only the Full-Dimensional $N_f = 20$ Feature Set Achieves the Arbitrary %C = 90% Benchmark.

authorized. By adjusting the threshold at which a device is successfully authorized, the TVR can be increased; however this could also increase the FVR.

An $N_d = 3$ class model was developed using the MDA/ML classifier for fingerprints generated with FPGA hardware. This model used a full-dimensional feature set of $N_f = 20$. The model was trained using $N_{Tng} = 500$ fingerprints per device. An additional $N_{Tst} = 500$ fingerprints per device were used to compare how much device 1 looks like device 1, device 2 looks like device 2, etc. These results are illustrated in Figure 4.9. The dashed line in Figure 4.9 represents an arbitrary benchmark of TVR>0.9.

The benchmark of TVR = 0.9 resulted in the corresponding FVR = 0.02 for devices $d = 1$ and $d = 2$. Classification performance for device $d = 3$ is perfect throughout this trial.

Using the same $N_d = 3$ class model and full-dimensional $N_f = 20$ feature set, devices $d = 4$ and $d = 5$ were introduced as rogue devices. These $N_d = 2$ additional devices were individually compared with every authorized device. The Rogue Accept Rate (RAR) is the rate at which rogue devices, posing as authorized devices, are incorrectly accepted as the claimed identity. The relationship between TVR and RAR is shown in Figure 4.10. The dashed black line in Figure 4.10 represents an arbitrary benchmark of $TVR > 0.9$. At this benchmark, nearly all cases achieve perfect classification performance of $RAR = 0.0$ and $TVR = 1.0$. Only the "4 looks like 3" case has imperfect performance of $RAR = 0.03$ at $TVR = 0.9$.

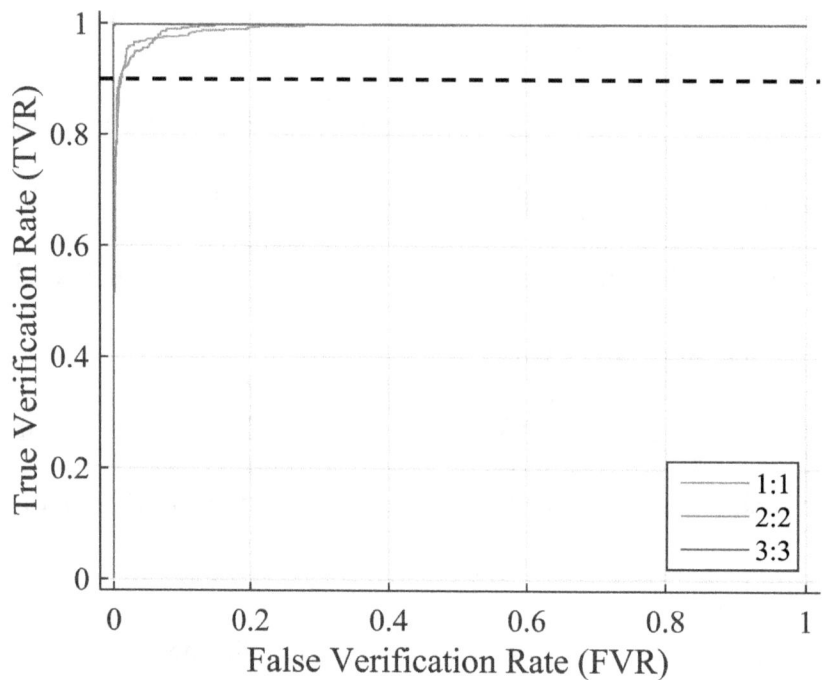

Figure 4.9: Full-Dimensional ($N_f = 20$) Authorized Device Verification Results for MDA/ML Model and Signals at $SNR_c \approx 55.0$ dB.

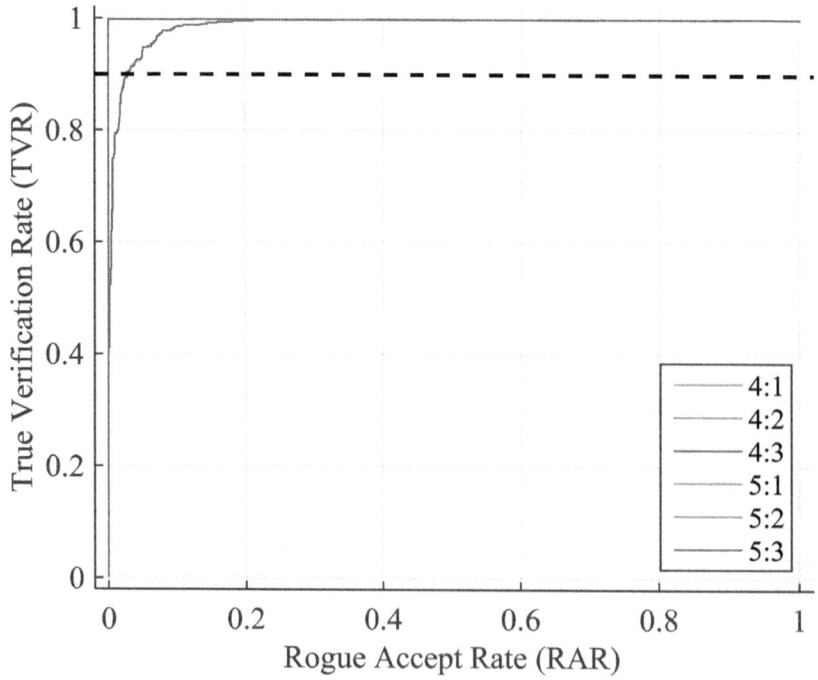

Figure 4.10: Full-Dimensional ($N_f = 20$) Rogue Device Rejection Results for MDA/ML Model and Signals at $SNR_c \approx 55.0$ dB.

The performance of qualitative DRA for Amp-only, Re-only and Im-only was statistically equivalent based on 95% CI. The performance of Phz-only was significantly lower than the other qualitative DRA feature subsets based on 95% CI. Therefore, Amp-only was used to generate a computationally light model for device ID verification.

The same verification process described previously was repeated for a reduced-dimension $N_f = 5$ feature set containing Amp-only fingerprints. An arbitrary benchmark of $TVR > 0.9$ and $FVR < 0.1$ is used for Authorized Device assessment in Figure 4.11a. Similarly, an arbitrary benchmark of $TVR > 0.9$ and $RAR < 0.1$ is used for Rogue Rejection assessment in Figure 4.11b. With this reduced-dimension $N_f = 5$ feature set, performance was significantly reduced. The FPGA implementation of the full-dimensional $N_f = 20$ fingerprint generator consumed only 17% of the total X310 FPGA resources. Implementing a less computationally complex model is not justified due to the poor performance results.

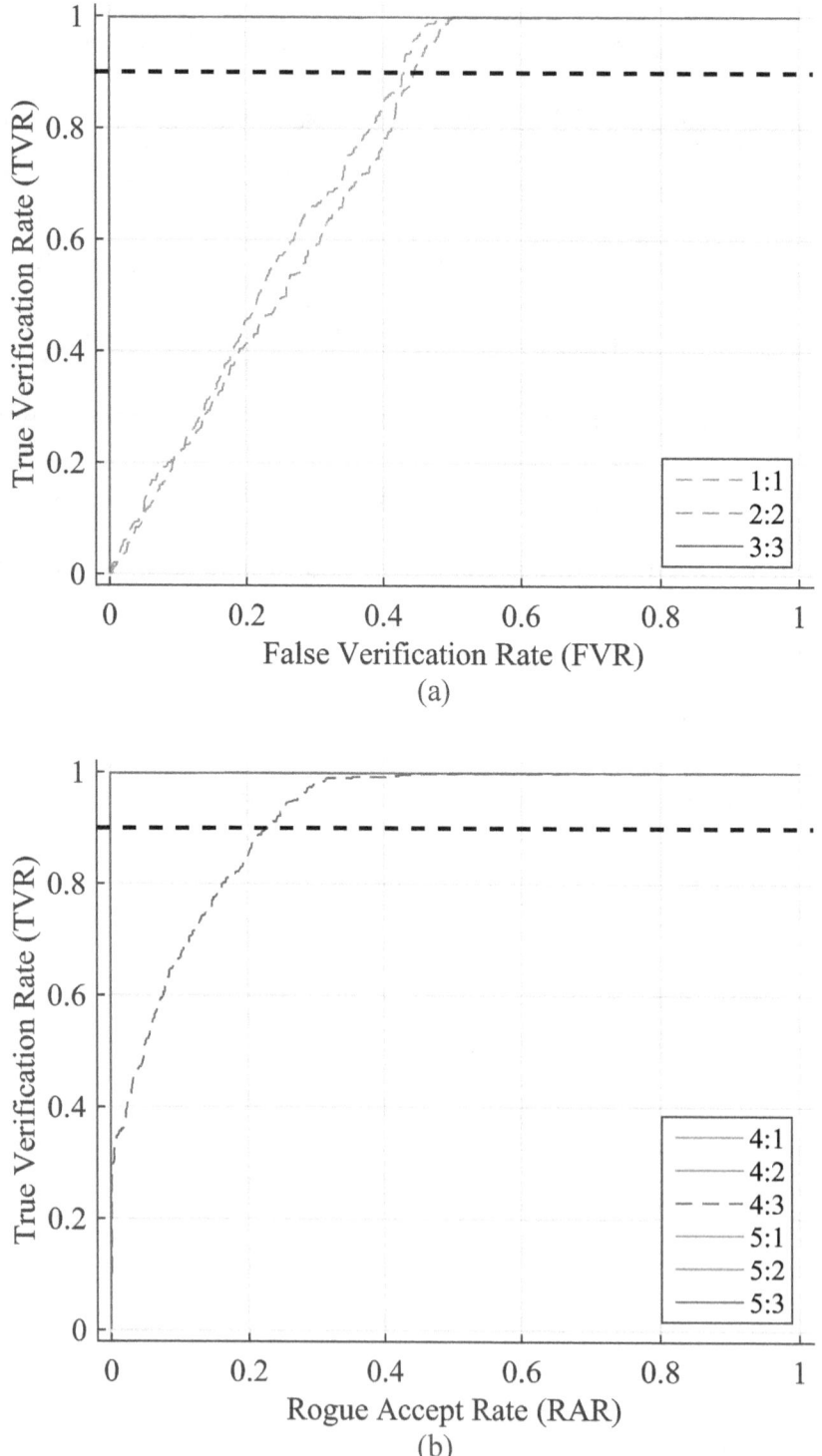

Figure 4.11: Authorized device and rogue rejection for Amp-Only DRA feature set for MDA/ML Model and Signals at $SNR_c \approx 55.0$ dB.

## 4.6 FPGA Performance

FPGA simulation software was used with the fingerprint generation design to determine the latency in generating fingerprints from a received ZigBee transmission. A ZigBee beacon request collected with the X310 in radio mode was input to the FPGA simulation of the fingerprint generation design so that the exact timing characteristics could be examined. Because the FPGA operates synchronously on a fixed clock, simulated timing performance will exactly match FPGA hardware timing performance. Figure 4.12 displays the fingerprint generation latency of the X310 FPGA design. Figure 4.12 shows the real-valued waveform of the ZigBee beacon request with the ROI (Region of Interest) and Payload highlighted. The vertical red line in Figure 4.12 indicates the point at which fingerprint generation is complete. As illustrated in Figure 4.12, fingerprint generation completes before most of the Zigbee payload has been received.

As shown in Figure 4.12, the latency time taken by the FPGA to generate a complete fingerprint of a single ZigBee beacon request is short in relation to the length of the payload. Because of the short latency in fingerprint generation time, in future work the payload could be selectively accepted or rejected based on fingerprinting results in real time, with no loss in Zigbee receiver data throughput. Figure 4.13 displays this fingerprint generation latency in greater detail.

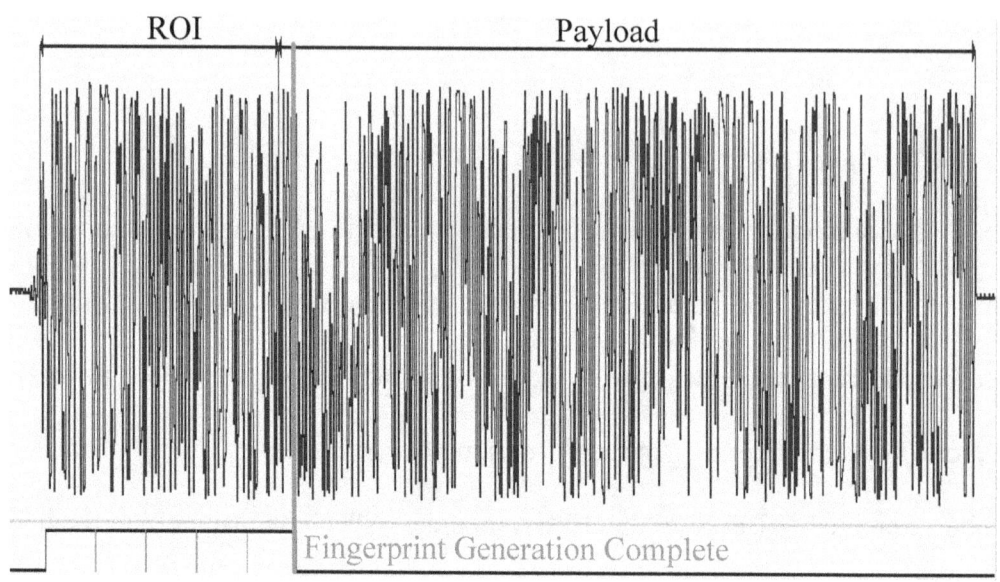

Figure 4.12: Processing Latency of X310 FPGA Real-Time Fingerprint Generation.

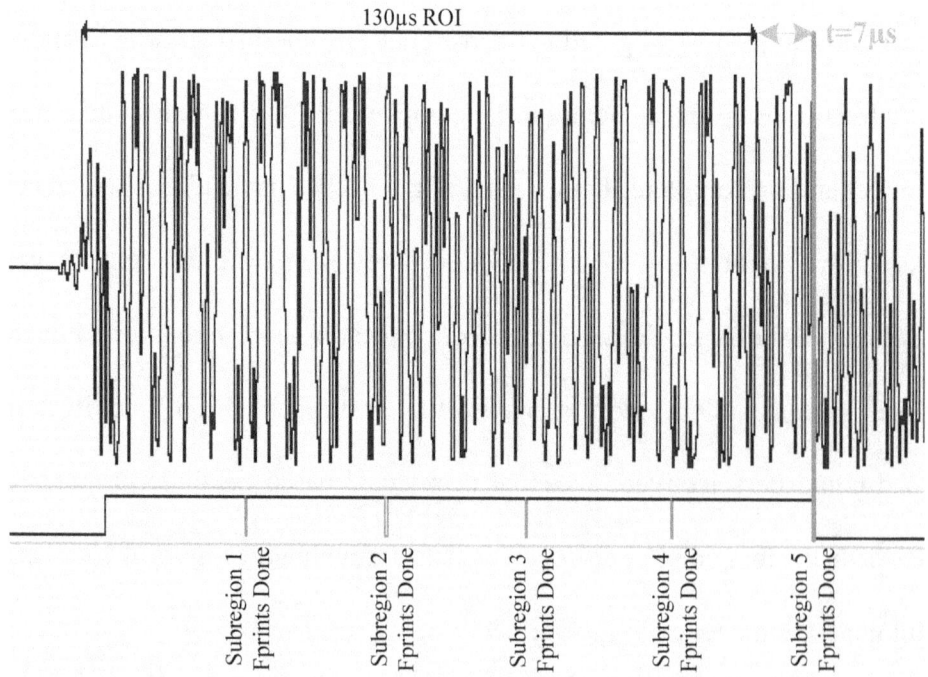

Figure 4.13: X310 FPGA Fingerprint Generation Processing Latency of $t = 7\mu s$ after Reception of ROI.

As shown in Figure 4.13, the complete $N_f = 20$ fingerprint is successfully generated at $t = 7\mu s$ after the ROI is received. The $t = 7\mu s$ latency of generating the fingerprint is minimal when compared with the $t = 370\mu s$ duration of the payload. Assuming a similarly short latency for device classification on FPGA hardware, the envisioned air monitor could be configured to selectively reject ZigBee transmissions from unauthorized devices in real-time.

# V. Summary and Conclusions

## 5.1 Research Summary

ZigBee networks are currently in place for functions such as monitoring medical devices, relaying electrical usage information to utility companies, and maintaining home automation systems. Due to the sensitive nature of many ZigBee applications, maintaining a high level of security is essential. Traditional security techniques for ZigBee networks are predominantly based on presenting and verifying device bit-level credentials (keys). While effective to some degree, bit-level-only security is becoming increasingly insufficient and ZigBee networks are vulnerable to attack by any unauthorized rogue device that can obtain and present bit-level credentials for an authorized device. Even without prior knowledge of the correct key, replay attacks can still be employed in which a packet transmitted by an authorized device is collected and later replayed by an unauthorized device [18].

Previous related research in [5,20,23] has shown that an additional Physical layer (PHY) of security can be applied using Radio Frequency Distinct Native Attribute (RF-DNA) fingerprinting to augment ZigBee bit-level security. RF-DNA exploitation involves generating a uniquely-identifiable "fingerprint" from PHY waveform features extracted from emissions of a particular device. The RF-DNA fingerprint is then used to discriminate devices from one another, even when identical (valid or false) bit-level credentials are presented. While previous AFIT research has demonstrated the effectiveness of MATLAB simulation-based RF-DNA classification of ZigBee devices, this research provides the next step towards achieving real-time device classification and verification. A complete RF-DNA based security solution for ZigBee devices in the form of an air monitor is proposed in [23]. The air monitor would be physically co-located with ZigBee devices and actively accept or reject signals from other ZigBee devices based on their fingerprint signatures. The purpose of this research was to demonstrate feasibility of the air monitor concept using an Ettus Research X310 Software-Defined Radio (SDR) hosting a Kintex-7 Field Programmable Gate Array (FPGA).

## 5.2 Findings and Contributions

An X310 hardware-based design was developed and evaluated in support of taking the next step towards achieving reliable air monitoring capability. The design and demonstration was based on a reduced-complexity MATLAB model of the traditional RF-DNA fingerprinting process.

The reduced-complexity MATLAB model included extraction of RF-DNA fingerprint features from ZigBee preamble responses. The preamble was divided into subregions, over which variance-only statistical features were calculated for the instantaneous Real ($Re[n]$), Imaginary ($Im[n]$), Phase ($\phi[n]$) and Amplitude ($a[n]$) time-domain responses. A full-dimensional fingerprint therefore contained a total of $N_f = 20$ features. MATLAB-generated fingerprints ($\vec{F}_M$) were created for each of $N_d = 3$ AVR RZUSBstick ZigBee devices and Multiple Discriminant Analysis/Maximum Likelihood (MDA/ML) classification performed. The average cross-class accuracy ($\%C$) for the initial trial using full-dimensional $\vec{F}_M$ fingerprints exceeded an arbitrary benchmark of $\%C > 90\%$ at a collected Signal-to-Noise Ratio ($SNR_c$) $SNR_c \approx 55.0$ dB.

An FPGA-compatible fingerprinting design was developed based on the MATLAB mode. The FPGA model was simulated using ModelSim simulation software. Experimentally collected ZigBee waveforms were collected using the X310 SDR which produced baseband sequence signals ($\{s_{BB}[n]\}$) that were inputs to the FPGA simulation to recreate actual operation. The FPGA Simulated ($\vec{F}_S$) fingerprints were created and classified using the same MDA/ML classifier. In this case, the arbitrary $\%C > 90\%$ benchmark was exceeded for an initial trial with $SNR_c \approx 55.0$ dB.

The simulated FPGA design was integrated into the X310 FPGA hardware by instantiating the fingerprint generator following the Digital Down-Conversion (DDC) on the X310 Kintex-7 FPGA. The FPGA-Hardware ($\vec{F}_H$) fingerprints were generated in real-time and streamed to the X310 interface computer. MDA/ML classification performance was assessed for a total of $N_{Trl} = 20$ independent experimental trials, the results of which consistently exceeded the arbitrary $\%C > 90\%$ benchmark.

Following full-dimensional ($N_f = 20$) assessments, Dimensional Reduction Analysis (DRA) was employed and classification performance evaluated using dimensionally reduced $N_f = 5$ feature sets. Using DRA, it was determined that fingerprints containing only on $Re[n]$, $Im[n]$, or $a[n]$ features produced statistically similar performance of $\%C \approx 66\%$, where statistical equivalence is based on 95% Confidence Intervals (CI) across $N_{Trl} = 20$ independent experimental trials. Fingerprints containing only $N_f = 5$ $\phi[n]$ features produced statistically poorer classification performance of $\%C \approx 50\%$ across the $N_{Trl} = 20$ independent experimental trials. While the full-dimensional $N_f = 20$ feature set consistently achieved the arbitrary $\%C > 90\%$ benchmark, none of the $N_f = 5$ DRA dimensionally reduced feature sets achieved the benchmark. As designed and implemented, the full-dimensional fingerprint generator only utilized 7% of the X310 Kintex-7 FPGA resources. Because of the low amount of FPGA resources required to implement the full-dimensional fingerprint generator, and the statistically poorer $\%C$ performance, the dimensionally reduced $N_f = 5$ fingerprint model is not justified in this application.

While the MDA/ML classification provided a "looks most like?" best match assessment, device ID verification was performed to conduct a "looks how much like?" assessment using the $N_d = 3$ authorized devices claiming to be themselves. A Euclidian distance measure of similarity $z_v$ was calculated and compared to a verification threshold value $t_v$ to make a binary accept-reject decision based on the device's claimed identity. By varying threshold $t_v$, the relationship between True Verification Rate (TVR) and False Verification Rate (FVR) was analyzed. TVR is the percentage of instances where a device is *correctly* granted network access after claiming its own identity. FVR is the percentage of instances where a device is *incorrectly* granted network access after claiming an identity that is not its own. For the $N_f = 20$ full-dimensional feature set, an arbitrary benchmark of $TVR > 0.9$ and $FVR < 0.1$ was achieved for all authorized devices. For $N_f = 5$ DRA feature sets one of the three devices achieved this benchmark while all other devices failed. For final proof-of-concept demonstration, two rogue devices were introduced and presented claimed IDs matching each of the authorized device IDs (a total of 6 rogue assessment scenarios). A comparison of TVR with Rogue Accept Rate (RAR) was made where RAR is the percentage of instances where a device is *incorrectly* granted network access after claiming the identity of an authorized device. For the full-dimensional case, an arbitrary benchmark of $TVR > 0.9$ and $RAR < 0.1$ was achieved for all 6 rogue scenarios.

Implementation of the full-dimensional fingerprint generator required only 7% of the Xilinx Kintex-7 FPGA resources. Therefore, even the mid-level Kintex-7 used in the X310 has plenty of room for expanding the air monitor's capability. The implemented design was able to generate a full-dimensional fingerprint in $t = 7$ μs after the end of the ZigBee preamble was detected. This latency is a small fraction of the $t = 370$ μs ZigBee payload duration. Because of the relatively short fingerprint generation latency, an unauthorized device transmission can easily be rejected by the air monitor in real-time with no loss in ZigBee data throughput from authorized devices.

## 5.3 Recommendations for Future Research

This research demonstrates that a low-cost SDR platform with an on-board FPGA is viable for air monitor implementation. As used here for initial proof-of-concept, the RZUSBstick ZigBee devices were successfully discriminated using RF-DNA fingerprints generated on the X310 SDR FPGA. The results here set the stage for additional hardware-oriented research avenues, including:

1. Increase RF-DNA Functionality on FPGA: This research is the first step towards a complete implementation of the air monitor. FPGA capabilities can be expanded by implementing additional RF-DNA functionality that was not

addressed here, including MDA model development and ML classification. Device ID verification can also be implemented on the FPGA hardware.

2. Consider Alternate Test Statistics: Traditional RF-DNA fingerprint generation involves generating a myriad of high-dimensional test statistics on a general-purpose computer having virtually unlimited time and computing resources. In migrating some of these more promising test statistics to a real-time FPGA implementation having fixed hardware resources, it will be necessary to consider which test statistics offer the best performance and which can be implemented within hardware resource constraints.

3. Perform a Real-Time Demonstration: The culmination of AFIT's wireless RF-DNA Fingerprinting research will be successful implementation and demonstration of an air monitor. This includes demonstrating that bit-level security can be augmented by PHY RF-DNA fingerprinting to provide enhanced security in real-time with enhanced speed, efficiency and robustness.

# Bibliography

[1] Atmel. RZUSBstick. http://www.atmel.com/tools/rzusbstick.aspx.

[2] H. Dawid and H. Meyr. CORDIC Algorithms and Architectures. Chapter 24 in Digital Signal Processing for Multimedia Systems, 1999.

[3] M. Donadio. CIC Filter Introduction. For Free Publication by Iowegian, 1999.

[4] C. Dubendorfer. Using RF-DNA Fingerprints to Discriminate ZigBee Devices in an Operational Environment. Master's thesis, Air Force Institute of Technology, March 2013.

[5] C. Dubendorfer, B. Ramsey, and M. Temple. An RF-DNA Verification Process for ZigBee Networks. In Military Communications Conference (MILCOM), pages 1–6, October 2012.

[6] Ettus Research. Bandwidth Capability of USRP Devices, 2015. http://www.ettus.com/kb/detail/usrp-bandwidth.

[7] Google Code. Killerbee, 2015. https://code.google.com/p/killerbee/.

[8] F. Harris, Multirate Signal Processing for Communication Systems. Upper Saddle River, N.J.: Prentice Hall PTR, 2004.

[9] IEEE Computer Society. IEEE Standard for Information Technology– Local and Metropolitan Area Networks– Specific Requirements– part 15.4: Low Rate Wireless Personal Area Networks (WPANs). IEEE Std 802.15.4-2011 (Revision of IEEE Std 802.15.4-2006), pages 1–314, September 2011.

[10] S. Killpack. RF-DNA Fingerprinting Applied to Commerical SatCom Devices. Master's of Science, Air Force Institute of Technology, March 2012.

[11] R. Klein. M. Temple and M. Mendenhall. Application of Wavelet-Based RF Fingerprinting to Enhance Wireless Network Security. Journal of Communications and Networks, December 2009.

[12] R. Klein, M. Temple and M. Mendenhall,. Sensitivity Analysis of Burst Detection and RF Fingerprinting Classification Performance. 2009 IEEE Int'l Conference on Communications (ICC), June 2009.

[13] R. Klein, M. Temple, M. Mendenhall and D. Reising. Sensitivity Analysis of Burst Detection and RF Fingerprinting Classification Performance. Proc of IEEE Int'l Conf on Communications (ICC09), June 2009.

[14] N. Lindlbauer. Application of FPGA's to Musical Gesture Communication and Processing. Masters thesis, Landshut University, November 1999.

[15] R. Lyons. Understanding Digital Signal Processing. Upper Saddle River, NJ: Prentice Hall PIR, 2004.

[16] M. Madki. Analysis And Implementation Of Multi-Rate Cascaded Integrator Comb (CIC) Interpolation & Decimation Filter With Compensation Filter For Software-Defined Radio. International Journal of Engineering Research & Technology (IJERT), July 2013.

[17] J. McGuire. Radio Frequency Distinct Native Attribute (RF-DNA) Fingerprinting Applied to Commercial SatCom Short Burst Data Modems. Master's thesis, Air Force Institute of Technology, March 2014.

[18] P. Radmand, M. Domingo, J. Singh, J. Arnedo, A. Talevski, S. Petersen, S. Carlsen. ZigBee/ZigBee PRO Security Assessment based on Compromised Cryptographic Keys. In 2010 Int'l Conference on P2P, Parallel, Grid, Cloud and Internet Computing.

[19] Ramsey. STE6000 RF Test Shielded Enclosure. http://www.ramseytest.com/product.php?pid=25.

[20] B. Ramsey. Improved Wireless Security through Physical Layer Protocol Manipulation and Radio Frequency Fingerprinting. Ph.D. dissertation, Air Force Institute of Technology, September 2014.

[21] B. Ramsey, B. Mullins, W. Lowder, and R. Speers. Sharpening the Stinger: Tuning Killerbee for Critical Infrastructure Warwalking. In Military Communications Conference (MILCOM), October 2014.

[22] B. Ramsey, T. Stubbs, B. Mullins, M. Temple, and M. Buckner. Wireless Critical Infrastructure Protection Using Low-Cost RF Fingerprinting Receivers. Int'l Journal of Critical Infrastructure Protection, December 2014.

[23] B. Ramsey, M. Temple, and B. Mullins. Phy Foundation for Multi-Factor ZigBee Node Authentication. In IEEE Global Communications Conference (GLOBECOM), pages 813–818, December 2012.

[24] D. Reising. Classifying Emissions from Global System for Mobile (GSM) Communication Devices using Radio Frequency (RF) Fingerprints. Master's thesis, Air Force Institute of Technology, March 2009.

[25] D. Reising. Exploitation of RF-DNA for Device Classification and Verification Using GRLVQI Processing. Ph.D. dissertation, Air Force Institute of Technology, December 2012.

[26] D. Reising and M. Temple. WiMAX Mobile Subscriber Verification Using Gabor-Based RF-DNA Fingerprints. 2012 IEEE Int'l Communications conf (ICC12), June 2012.

[27] D. Reising, M. Temple and M. Oxley. Gabor-Based RF-DNA Fingerprinting for Classifying 802.16e WiMAX Mobile Subscribers. 2012 IEEE Int'l Conf on Computing, Networking & Communications (ICNC),  January 2012.

[28] G. Smart. Software-Defined Radio for Cognitive Wireless Sensor Systems. University College London, March 2010.

[29] W. Suski, M. Temple, M. Mendenhall and R. Mills. Using Spectral Fingerprints to Improve Wireless Network Security. Proc of IEEE Global Communications Conf (GLOBECOM08), March 2008.

[29] W. Suski, M. Temple, M. Mendenhall and R. Mills. Using Spectral Fingerprints to Improve Wireless Network Security. IEEE Global Communications (GLOBECOM), October 2008.

[30] A. Tao. Application of Application of Radio Frequency Distinct Native Attribute (RF-DNA) Fingerprinting to Commercial Satcom Short Burst Data Modems on Software-Defined Radio. Master's thesis, Air Force Institute of Technology, March 2015.

[31] S. Theodoridis, and K. Koutoumbas. Pattern Recognition. Academic Press, 4th Edition, 2009.

[32] M. Williams. Application of RF-DNA Fingerprinting to Improve WiMAX Security. Master's of Science, Air Force Institute of Technology, March 2011.

[33] M. Williams, S. Munns, M. Temple and M. Mendenhall. RF-DNA Fingerprinting for Airport WiMax Communications Security. Proc of 4th Int'l Conf on Net and Sys Security (NS28), September 2010.

[34] M. Williams, M. Temple and D. Reising. Augmenting Bit-Level network Security Using Physical Layer RF-DNA Fingerprinting. Proc of IEEE Global Communications Conf (GLOBECOM10), December 2010.